创意产业与自发性城市更新

CREATIVE INDUSTRIES
AND BOTTOM-UP URBAN REGENERATION

许凯 孙彤宇 著
Xu Kai Sun Tongyu

中国建筑工业出版社

图 1 厦门沙坡尾航拍照片
Fig. 1 Aerial View of Sha Po Wei in City of Xiamen

关于本书

创意改变城市。那么，在城市中，是什么让创意得以发生和发展？

本书关注一种城市现象。创意产业嵌入城市，与所在城区既有的功能与空间紧密关联，形成创意社区。这个现象揭示了产业推动下的、自下而上的城市更新模式是可行的，且能创造一定的城市特色。基于自组织理论，本研究认为创意社区中以产业为核心的功能系统是一个开放系统，它的发展是由系统中的各个利益相关者之间的相互作用推动的，社区的空间形态也必然显示出自组织的相应特征。本书向读者介绍近些年中国城市化过程中出现的那些自发生长的、自由自在的和美丽的创意社区，并向大家揭示，什么是隐藏在空间背后的力量。

通过分析这些创意社区，本书希望传达一种观点，即"生长"是城市的天性。城市里的各个利益相关者都有权利改变和创造与他们息息相关的城市空间。我们还试图回答，"规划"作为我们用来调控空间的工具，它的作用是什么，以及它如何与"自下而上"的城市发展力量合作，来共同创造更美好的城市。

About This Book

Creativity changes cities. In cities, what inspires the creativity and encourages its development?

This book focuses on an urban phenomenon. Creative industries embed in the city and establish close association with existing function and space of the city areas where they are situated, thus creative community is formed. Such urban phenomenon showcases that bottom–up regeneration driven by creative industries is an applicable model for our urban future. Based on theory of self–organization, this book argues that the function system of the creative community, in which creative industry plays a key role, is a complex system, whose development is driven by interactions between stakeholders in the system. Its urban form therefore reflects rules of self–organization correspondently. This book introduces to readers those spontaneous, free and beautiful creative communities that have emerged in the process of urbanization in China in recent years, and reveals what is the power behind space.

By analyzing the creative community, the book hopes to convey the idea that the city always evolves. All stakeholders in the city should have the right to change and create urban spaces that are relevant to them. The book also attempts to answer, as a tool of spatial control, what is the role of "urban planning", and how can it work together with "bottom–up" forces, in order to create better cities.

前言

　　城市是一个开放的复杂系统，因而必然具有自组织发展的特点。如何理解这种特点，以及如何运用城市政策和管理手段引导和规范它的发展，是城市管理者面临的重要挑战。

　　城市形态的形成除了与自然环境、地理条件、经济条件、技术条件及人口数量等要素相关以外，更与一定时期的社会体制、文化特性、地方习俗以及相关城区居民、企业、投资者等的利益诉求产生错综复杂的关联。城市规划以假想的使用者的视角来进行物质空间形态规划与资源配置，确定了一个城市系统的初始状态。而当城市初步形成，有了真实的使用者之后，城市的物质空间之上便形成了真实的经济活动和社会活动。城市的实际使用者不断变化的需求，将对城市物质空间形式、基础设施容量等物质性要素发出求变的信号。因此，对城市更新的需求从城市一经形成便已出现。调整和更新是城市发展的常态，我们可以从城市建成区的很多自发性更新案例看到那种汹涌的推动城市发展的源动力。然而城市管理部门由于一直坚守城市初始设定的各种规则，对于这样的现象显得束手无策，常常陷入所谓"一管就死，不管就乱"的局面。

　　城市发展过程中对自身的不断调整和完善，是城市这个复杂系统内各种要素相互竞争与协同以达成动态平衡的过程。我们可以设想，如果能够有效引导和利用这种强大的城市发展动力，城市将会朝着高效和良性的方向不断发展。在利益相关者博弈过程中，城市管理者必须成为积极参与者，而不应该是置身于局外的旁观者，否则利益相关者的诉求就难以通过法定的手段得以实现。城市管理者需要建立利益相关者之间的谈判机制，帮助他们在扩大自身利益或至少确保利益不受损失的前提下达成相互妥协，形成共赢局面。我们的城市治理手段也需要有更多的弹性。由此自发更新得到更好的引导，进一步成为城市发展的创新动力。

　　本书记录了许多典型的城市自发更新案例，从自组织理论的角度剖析其中的利益相关者如何为提升所在城区的经济和文化特色所做出的贡献，以及这种自发更新所形成的活色生香的城市物质空间形态和城市生活形态。通过这项研究，我们揭示了推动城市可持续发展的底层强大动力，并呼吁城市管理者应积极发挥作用，创造城市治理的弹性和创新机制，对自发的城市更新进行良好的引导和规范，并给予资源上的支持。和谐的、创新的城市，必然由此而发生。

2018年11月4日于同济大学UrbanLab

PREFACE

City is an open and complex system with the characteristics of self–organization. How to understand these characteristics and how to establish urban policies and management tools to guide and regulate this self–organized process, are important challenges for city managers.

The formation of urban form is not only related to the environmental, geographical, economic, technical conditions and the number of population, but also related to the social system, cultural characteristics, local customs and the interests of the relevant urban residents, enterprises and investors. Urban planning determines the initial state of a city system by making use of hypothetical users to carry out the planning of physical space and the allocation of resources. However, once the initial formation of the city is completed, and with the emergency of specific users, the physical space becomes places for real economic and social activities. The changing demand of users of the city will send out directives for the change of material elements such as modified urban form and added infrastructure. In this sense, the demand for urban regeneration emerges ever since the formation of the city is completed. We can see from many cases the surge of such power behind them.

The continuous adjustment and perfection in the process of urban development is a process in which various stakeholders compete and cooperate with each other towards maximized common interest. We can imagine that if we can effectively guide and make use of this inner power of urban process, the city can grow with both efficiency and harmony. In such process, when various stakeholders interact, urban managers will become active participants rather than bystanders. A negotiation mechanism shall be established to help stakeholders reach mutual compromises, on the premise of maximized common interests or at least ensuring that they do not suffer losses. Approach to urban governance also needs to be more resilient.

This book documents many cases of spontaneous urban regeneration, and analyzes how stakeholders contribute to enhancing the economic and cultural characteristics of their urban areas from the perspective of self–organization theory, as well as the vivid and fresh urban forms and urban lives brought by these spontaneous regeneration processes. Through this research,we will reveal the bottom–up stimulators behind city's sustainable growth, as well as call on urban managers to play an active role in setting up innovative and resilient mechanism of urban governance, which can guide and regulate the spontaneous urban regeneration, while not interrupting those vivid interaction between stakeholders. Harmonious and innovative cities are bound to arise from this.

Prof. SUN Tongyu

Urban Lab, Tongji University, Shanghai

研究团队
RESEARCH TEAM

研究主持
RESEARCH LEADERS

许凯 ｜ 在维也纳工业大学建筑学院获得博士学位，师从克劳斯·泽姆斯罗特（Klaus Semsroth）教授。现任同济大学建筑与城市规划学院城市设计团队副教授。主要研究领域是城市设计与产业空间规划。他是"尤根设计"和 Urban Lab 的主要创办者，积极从事建筑设计和城市设计方面的工作。

Dr. Xu Kai is an associate professor in Tongji University of Shanghai. He obtained his doctor degree in Vienna University of Technology. Prof. Klaus Semsroth was his tutor professor. His research areas include urban design and industrial space planning. Being the co-founder of Jugend Architecture and Urban Lab, he also actively practice in architecture design and urban design.

孙彤宇 ｜ 同济大学建筑与城市规划学院博士，德国柏林工业大学和斯图加特大学访问学者。现任同济大学建筑与城市规划学院副院长、教授、博士生导师，中国建筑学会城市设计分会常务理事，建筑教育评估分会副理事长，中国建筑学会标准工作委员会委员，上海市绿色建筑协会副会长。主要研究领域为城市设计及建筑设计理论与方法，主持多项国家和省部级科研项目。同时他也是活跃的从业建筑师，担任同济大学建筑设计研究院（集团）有限公司都市建筑设计院四所所长和主创建筑师，有多项设计获国家和地方建筑设计奖。

Prof. Dr. Sun Tongyu obtained his Ph. D. degree in Tongji University in Shanghai. He was the visiting scholar in TU Berlin in 2003, and in Stuttgart University in 2008. He is now the vice dean of the College of Architecture and Urban Planning in Tongji University, executive director of urban design branch, vice chairman of architecture education evaluation branch, member of code constitution branch of Architectural Society of China, he is also the executive director of green building society of Shanghai. His main research fileds are urban design and architecture design. He leads various national and provincial level research projects, and also actively practices in architecture design and urban design. He is the director of design division No. 4 in Tongji Design Group.

莫拉登·亚德里奇（Mladen Jadric） ｜ 奥地利维也纳建筑师，任教于维也纳工业大学。师从阿尔索普教授，维也纳工业大学博士。维也纳工业大学中奥交流项目负责人，奥地利艺术家协会成员，奥地利建筑师协会副主席。

Mladen Jadric, teaching and practicing architecture in vienna, Austria. His dissertation was mentored by Prof. William Alsop at TU Vienna. He is head of exchange programs and workshops with Asia and double-degree program with Tongji University, Shanghai, China. He is member of Vienna Künstlerhaus, and vice chairman of Chamber of Architect of Austria.

克劳斯·泽姆斯罗特 （**Klaus Semsroth**） | 奥地利维也纳工业大学教授、博士。2000 年至 2013 年任维也纳工业大学建筑与空间规划学院院长。曾担任奥地利规划与区域发展协会主席，奥地利历史遗迹保护协会顾问团成员，维也纳市房地产协会顾问团成员。主要研究领域是城市规划、城市设计和城市发展历史。他是《卡米洛·西特全集》丛书的主编与学术负责人。2002 年，他与莫拉登·亚德里奇一起发起和建立了中奥建筑院校的学术合作。

Prof. Klaus Semsroth obtained his Ph. D. degree and habilitation (Univ.Docent) from Vienna University of Technology. From 2000 to 2013 he was the dean of the Faculty of Architecture and Spatial Planning, TU Vienna. He was chairman of an Austrian organization of urban and regional planning, member of advisory board for preservation of historical monument as well as member for advisory board for real estate of Municipality of Vienna City. His main research scopes are in urban planning and urban design，as well as in the field of urban history. he is the main editor and director of *The complete edition of Camillo Sitte*. Together with Mladen Jadric, he is the initiator of cooperation between TU Vienna and China since 2002. He is practicing architect since 1985.

研究参与人员
RESEARCH PARTICIPANTS

Jung Jea Hyun，Katerina Gurova，Luz Pardo del Viejo，Soya，Linda Sköneskog，Lee Su Hye，Lina Lim，Jung Jea Hyun，Marie Simon，Sara Willaert，白璐 ，陈梦梦，杜叶诚，戴子钰 ，金刚 ，李蔓竹 ，罗愫 ，刘亚飞，王登恒，吴繁文 , 杨舒丹，叶磊，赵畅 ，张家洋 ，张黎晴，张起铭，朱薛景，赵月僮。

目录

CATALOGUE

1 创意社区的形成，
一种由"自下而上"力量推动的城市进程？

THE FORMATION OF URBAN CREATIVE COMMUNITY,
AN URBAN PROCESS DRIVEN BY BOTTOM–UP POWER?

图 2 维也纳 WUK 的内院
Fig. 2 Courtyard In WUK of Vienna

维也纳的 WUK 的内院是一个"场所",这里不定期举办各种艺术活动,吸引着艺术家、市民和游客的参与。

The courtyard of WUK is a "place", where various art events take place from time to time. This is a core attraction of WUK to bring artists, citizens and tourists together.

图 3 维也纳 WUK 内院里的艺术活动
Fig. 3 Art Events in WUK of Vienna

图 4 WUK 里的仓库空间被转换成服务艺术活动的空间
Fig. 4 Converted Warehouse Space in WUK for Art Activities

创意社区是与某种特殊城区紧密关联的,这些城区不仅容纳了城市住宅和服务业功能,它们也是创意产业的所在地。有一种特殊的人群,他们是创意产业的从业人员,他们使用着这里的服务设施和空间,他们中的一些就住在这里或者附近的城区里。产业和城市其他功能的极致融合,令这些城区无一例外地体现出强烈的"社区"的特征。

很多创意社区并不是通过有组织的"规划"形成的,而是在既有城区的基础上,创意产业逐步置换既有的建筑功能、改造既有的城市空间,慢慢形成的。这样的例子在欧洲和美国城市化程度较高的地方屡见不鲜。奥地利维也纳的中心城区,创意产业占用一些城市建筑和废弃的基础设施,发展出富有特色的社区。其中著名的是 WUK,艺术家聚集在这个市中心的废弃仓库里,将它改造成一处艺术聚落。它不仅容纳艺术家和他们的创作活动,更把这里变成一处"场所",城市事件在这里频繁发生着,吸引着无数市民和游客的参与(图 1,2,3)。Stadtbogen 是第一次世界大战以前建造的城市铁路桥,它的桥洞空间今天成为艺术工作室、作坊和特色酒吧的聚集地,建筑遗产和新兴的产业功能在这里相遇(图 4,5)。在意大利米兰的 Zona Tortona 区,时尚、设计和艺术产业慢慢改造被废弃的产业区,将这个区域发展成一个新兴的特色城区(图 6,7)。在德国,一些大的城市更新计划覆盖下的城市中心区域(如杜塞尔多夫的媒体港、汉堡的仓储城等),也有很多地块是通过"自下而上"的方式发展起来的。

本书关注中国城市里自发形成的创意社区。在中国城市发展的现阶段,房地产推动下的城市再开发几乎成为城市更新的唯一方式。为了追求更高的经济利益,原有城市结构被完全替换,以实现更高的开发强度并符合新的社会人群需要。尊重和延续原有物质和社会结构的可持续城市发展,在这种模式下是很难实现的。尽管如此,近年来有些城市里出现了一些特殊城区,它们在没有外来开发主体的情况下,通过引入创意产业,在基本不改变原有城市结构的情况下进行"自下而上"的完善更新。这些城区都几乎无一例外地成为城市里具有特色的区域。上海的田子坊、M50,深圳的大芬村,厦门的沙坡尾和曾厝垵,北京的 798 艺术区、草场地和宋庄,都是这方面的杰出例子(图8,9,10)。该现象揭示了,创意产业驱动下的、"自下而上"的城市更新模式是可行的,而且能够为城市带来了新的发展和特色。

我们必须指出,中国的创意社区,是在与国外案例很不相同的外部条件下成长起来的。其中的一些案例,发生在城市扩张中出现的特殊区域(例如城中村或者转型过程中的工业区),城市规划和城市管理的缺位却让自发更新得以发生。另一些案例,则坐落于带有历史特征的城市中心区,以原住居民和创意企业为主体的自发更新需要面对来自规划部门、城管部门和开发商的重重阻力,在拆了又造、造了又拆的情况下顽强地发展,直至获得社会认同而得以合法化。这些特殊的外部条件和发展历程,不仅给予这些中国案例独特的研究视角,更是提出一个核心问题,那就是,在城市更新中,"自上而下"的空间调控工具(如城市规划)是不是必不可少的?如果是的,那么它和"自下而上"的发展动力之间的恰当关系是什么?

我们还需要指出,创意城区的形成本质上是一种产业推动的、"自下而上"的城市发展进程,该进程和当代城市发展中盛行的 "从图纸到空间"的转译进程是很不同的。它为何发生,如何发展,最终导向什么,都是我们很不熟悉的内容。这导致了城市规划和管理部门其实并不知道如何面对这样的城市进程,以至于在一些自发更新发生的时候(有些甚至已经进入良性发展的阶段)惊慌失措,做出一些不恰当的举措,令城市痛失一次次发展创意的机会 [i]。

本书对上述现象的研究将聚焦于城市空间形态方面。研究将揭示创意城区的空间形态发展规律及其成因。通过对空间形态的分析,我们也会揭示,作为"自上而下"空间工具的规划和相应的政策、管理,在自发城市更新中扮演的角色是什么,以及它们能否就自发更新的规律做出调整,其效果如何等。

研究将依据下面的框架展开。本章介绍了研究的基本背景。第二章和第三章分别介绍当代创意产业的定义和特点,以及作为本研究基石的"自组织理论",该理论不仅对当代的自然和社会科学产生了很大的影响,也为我们认识城市发展规律提供了一种根本性的视角。运用该理论,我们尝试揭示出创意城区自发展的动力机制和空间特征,这部分研究见于本书的第四章。在本书的第五章,我们将通过一系列分析向读者展示国内五个创意社区,让大家了解现象之后的特征和规律。本书的第六章,探讨城市规划作为"自上而下"的空间调控手段,能否在自发更新中起到作用。当然,这很难通过实证研究来揭示(迄今为止这样的实际案例非常少),只能够运用虚拟的方式。我们展示了本书作者作为导师带领同济大学 2016 和 2017 两届研究生同学做的城市设计课题。第七章,我们提出了"规划不可规划之事" [ii] 的论点,并提出相应的策略,作为研究的结论。

图 5 维也纳 Stadtbogen 桥洞下入驻的小型产业
Fig. 5 Small Industries Beneath The Railway Bridge in Stadtbogen of Vienna

创意产业的集聚往往会选择出人意料的城市区域,它们所表现出来的空间形式也同样出人意料。

Creative industries tend to agglomerate in unexpected urban areas, and their spatial forms are equally unexpected.

图 6 桥洞空间里的特色餐厅
Fig. 6 Special Restaurant Under TheRailway Bridge

桥洞空间成为餐厅、酒吧和工作室,说明创意产业对既有空间的极大适应能力。

The space beneath the railway bridge becomes a place for restaurants, bars and workshops, which shows that how flexibly the creative industries can be adapted to existing space.

图 7 米兰的 Zona Tortona 区里被创意产业激活的地块
Fig. 7 Urban Block in Milan's Zona Tortona Which is Activated by Creative Industries

与时尚、设计和媒体相关的产业进入既有的工业区，巧妙地转换街坊里的建筑功能。而外部的空地往往被转变成使用效率很高的公共空间，适应人流聚集、展会、餐饮和艺术活动的需要。

Fashion, design and media industries embed into former idle industrial area and replace the former function. Spaces between buildings are converted into public spaces that fit the demand of agglomeration of people, exhibition and art activities.

图 8 米兰的 Zona Tortona 区里的某多功能特色空间
Fig. 8 A Multi-Functional Space in Zona Tortona

多种多样的产业内容可以混合在同一个空间里。上图的这个空间既有餐厅和咖啡厅的功能，也售卖家具和旅游纪念品，它又是一个"作坊"，为创意工作提供空间。

Various industries can stay in the same space. In this picture, we see a restaurant & bar can also be shop for furniture and souvenirs. It acts also as a workshop in which creative works find their place.

This book draws attention on a specific type of urban area, which not only contains urban housing and related service functions, but also is the home for creative industries. In addition to original inhabitants and people who work in related service, there is a considerable amount of people who work in creative industries. Accordingly, these people live in the same area where they work, or in nearby urban areas. Such urban areas have strong characteristics of being a "community". In this book, we call them "creative communities".

Many creative communities do not take shape through "planning" but through "bottom-up" process, in which, on the basis of existing urban context, creative industries gradually replace existing building functions and transform existing urban spaces until the creative communities are formed. Examples of such examples come into sight in many European and American cities. In Vienna, creative industries occupy former industrial buildings and abandoned infrastructure. One of the famous examples is WUK, where historic warehouse became home for artists, and the central courtyard becomes a "Place" for frequent events, which attracts numerous citizens and tourists (Fig. 1,2,3). And in the underneath place of the former railway bridge named Stadtbogen, fashion workshops and bars find their places. Cultural heritages and newly rising creative industries meet each other consequently (Fig. 4,5). In Milan's Zona Tortona (Fig. 6,7), fashion, design and art industries slowly transform abandoned industrial districts into Milan's most attractive creative area. In Ruhr Area, Germany, many urban blocks covered by the large urban renewal schemes are also cast through similar process, such as those in the Media Harbor in Dusseldorf and in the Speicherstadt in Hamburg, etc.

On this background, this book focuses on creative communities in China. At the present stage of urbanization in China, urban redevelopment driven by real estate investment has almost become the only way of urban renewal. In order to pursue higher economic benefits, the original urban structure has been completely replaced, in order to meet the higher development intensity (FAR) and meet the needs of new social groups. Sustainable urban development that respects the original spatial and social structures is difficult to be achieved in this model. Nevertheless, in recent years, some cities have seen the emergence of special urban areas that, without the introduction of external development funds, have been regenerated materially and socially through "bottom-up" development driven by creative industries. Moreover, almost without exception, these urban areas have become most characteristic areas of the city. Tian Zi Fang in Shanghai, M50 in Shenzhen, Da Fen Village in Shenzhen, Sha Po Wei and Zeng Cuo An in Xiamen, 798, Cao Chang Di and Somg Village in Beijing are outstanding examples among them (Fig. 8,9,10). These cases reveal that the "bottom-up" urban regeneration model driven by creative industries is an optional model for the city, which can bring new growth points and characteristics for urban development.

Creative communities in China have come into being in circumstances different from that in other countries. Some of these communities took place in specific area aroused during city's spatial expansion, such as in converting industrial areas or reputed "urban villages". To a certain extent, the absence of urban planning and strict urban administration in such areas gave ideal conditions for bottom-up regeneration. The other communities may take launch in central areas in cities, many of which have historical contexts. The regeneration led by original residents and creative enterprises has to face up to resistance caused by urban planning authorities, administrations and developers. Therefore, demolishing and rebuilding are most common phenomenon here, until the bottom-up regeneration is accepted and legalized by authorities. The specific circumstance mentioned here, gives a new and unique dimension to research on these cases. Besides, it leads to a core academic question of this book, which is, in urban regeneration, is a top-down spatial instrument (such as urban planning) indispensable? If so, then the next question is what is the nexus between it and the bottom-up development impetus?

We also need to point out, that the formation of creative community is a bottom-up urban process driven by industries. This process is different than the prevailing urban process in modern cities, namely a "metaphrase of drawings into space". Why and how the bottom-up process performs and where it leads are unknown to the planning and administration authorities. When it occurs, counter measurements are often made in a panic, which is inappropriate and, as a result, makes the city lose chances of redevelopment [i].

The research in this book is to take the spatial form of creative communities as the focus. The discipline of evolution and its underlying cause is to be revealed. Through the research on spatial form, we will also find out, what is the role of the top-down instrument, such as urban planning and correspondent spatial policy and administration, in the bottom-up regeneration process.

The research will follow this structure. In brief, after the introduction of research background in this chapter, chapter 2 is a presentation of creative industry, creative class and their intervention on space. Chapter 3 introduces the theory of self-organization, the cornerstone of this book and, furthermore, its influence on the research on modern industries and urban space. We will continue to employ the discipline of self-organization to unveil the dynamic mechanism and spatial characteristics of creative communities in chapter 4. Empirical research on 5 communities, supported by illustrations and analysis is to be demonstrated in chapter 5. Chapter 6 initiates a discussion on how the top-down instrument functions in bottom-up process. Certainly, it is difficult to display with an empirical method because of the rarity of actual cases. Nevertheless, we will show works from 2 master degree design studios offered by Tongji University of Shanghai, tutored by Prof. SUN Tongyu and Prof. XU Kai. In Chapter 7, we put forward the argument of "To plan the unplannable" [ii], and propose respective strategies as the conclusion of the study.

图 9 城中村里发展出来的创意社区：大芬村
Fig 9. Creative community developed from urban village: Da Fen Viallge

图 10 里弄住宅里发展出来的创意社区：田子坊
Fig 10. Creative community developed from Li Long housing area: Tian Zi Fang

图 11 渔村发展出来的创意社区：曾厝垵
Fig 11. Creative community developed from former fishing village: Zeng Cuo An

2 创意产业、创意阶层和他们需要的空间
CREATIVE INDUSTRY, CREATIVE CLASS AND THEIR SPACE

我们首先来了解一下创意社区的主角——创意产业。英国文化部将创意产业定义为"那些根源于个人创造力、技术和天赋的产业，它们通过产生和开拓知识产权创造财富和价值"，包含 9 个门类：①广告和市场营销 ②建筑 ③手工艺 ④设计(产品、平面、时尚) ⑤电影、电视和摄影 ⑥IT、软件和电脑服务 ⑦出版 ⑧博物馆、画廊和图书馆 ⑨音乐、表演和视觉艺术（DCMS，2001）。有一些研究者认为还应该增加游戏和科学技术里的 R&D 部门（豪金斯，2001）和教育产业（赫斯蒙德哈尔什，2002）。佛罗里达指出，人类的创新力是终极的经济资源（佛罗里达，2002）。也有学者指出，21 世纪的产业将越来越依赖于以创新为基础的知识生产能力（兰德里和比亚恩奇尼，1995）。创意产业已经被视为经济繁荣不可或缺的部分。

这是社会中一种特殊的人群，由于他们从事的是创新性工作，他们某种程度上具有相似的性格特征和需求。本研究所关注的创意社区，就是创意人群工作、生活和进行休闲活动的城区。这里特有的"自发性城市更新"，也是以他们为主体推动的。

他们和空间的关系是什么？他们需求的空间是什么样的？回答该问题可以解释，为什么创意企业和人群入驻城区之后，会对城区的空间进行持续不断的推动和改善，令自发性城市更新得以发生。从建筑内部环境的改变、对建筑物本身的改造，直至建筑外部环境的改善，空间更新无处不在，其幅度之大，是其他类型的办公区所没有的。不能不说，这和创意企业以及人群的内在特点有必然的联系。

首先，创意企业对小型化的工作空间有着持续的、巨大的兴趣。这与企业的规模有很大的关系。在全球化的趋势下，产业划分越发精细化，企业规模越来越小。这些小型的企业往往在全球化的产业链中占据一环，依靠创新来维持它的活力和竞争力，而不去按照传统方式做大做全产业链以寻求规模扩展。在很多国家里，小微企业（SME）[iii]无论在人数上还是企业数量上都相对大型企业占据压倒性的优势，他们在国家的经济中起着巨大的作用。尽管创意产业和小微企业的定义当时遵循的是不同的划分逻辑，但不可否认的是，大部分创意企业都是小微企业。它们在创业的初期往往无法负担很大的办公空间，却对空间的独立性有很高的需求。它们也非常在意空间的灵活度，因为企业的规模随时会因为业绩和发展思路的不同而变化，空间则需要对这些变化进行适应。因此，城中村、里弄住宅这些类型就因为能够满足这方面的需要而成为首选。而像 798 这样的案例，房屋空间其实并不小，但是提供了灵活划分成较小单元的可能性。

其次，对于很多创意企业来说，工作空间是展示他们产品的重要场所，或者在某种情况下，空间成为它们品牌形象的一部分。对一些类型的企业（如设计），工作空间本身就是他们产品的一部分。这会驱使他们首先选择有空间特色的城区来入驻，并在建筑的内部或外部空间改造方面投入很多的精力和创新力。在上海八号桥和 M50 艺术区，厂房空间被划分成小的单元，单元内部的空间设计多姿多彩（图 11、图 12、图 13）。这些单元往往把自己的外墙变成透明的玻璃橱窗，让过往的游客或其他的创意人群可以看见单元内部的空间和人的工作。上海田子坊的文化历史特色，也就自然而然地成为入驻企业的特色，而他们对空间的积极改造，也自然成为新田子坊特殊气质的一部分。

最后，创意企业能从"个体行动"走向"共同行动"，自发地推动公共空间的生成和生长，这是创意社区最令人振奋和赞叹之处。生产新的知识，是创意产业的基本职能，正是该职能令创意产业区别于其他产业。新知识的生产往往是来自于不同产业个体（企业或个人）之间的相互联系和合作，而个体之间产生联系所依赖的物质空间，就是公共空间。也只有公共空间的存在，才让创意社区作为"社区"的特质真正得以形成。从这个角度，我们就可以理解为什么创意企业对空间的自主改造往往会演变成一种"群体营造"，而"群体营造"的阵地就是公共空间。八号桥的公共展示活动空间、田子坊的街巷和沙坡尾的户外艺术空间，都是"群体营造"的结果（图14、图15、图16）。

图 12 上海八号桥的一处办公空间

Fig. 12 A working space in Bridge 8 of Shanghai

小型化的、共享的工作空间是创意企业青睐的空间形式。在这个例子里，通过玻璃外墙，周边办公室和经过的游客也能够看到企业工作的场景，进一步扩大"共享"。

Small and shared working space is preferable form of creative enterprises. In this case, through external wall made of transparent glass, neighboring office and passing-by tourists can see the sceneries of working.

图 13 上海八号桥的另一处办公空间

Fig. 13 Another working space in Bridge 8 of Shanghai

半开放的空间是创意企业工作空间常用的形式。在这个例子里，阳台是休闲的空间，从这里员工可以看到外部公共空间发生的活动。当然，在公共空间里活动的人群也可以看到他们，或者进一步通过很大的玻璃橱窗看到办公室里的情况。

Semi-open space is a prevailing form of working space of creative enterprises. In this case, the balcony is a leisurely place, from which the employees can watch the activities taking place in the public space. Certainly, the people in the public space can also watch them back, or, through the big glass wall, watch the working sceneries in the office.

In this chapter, we will get to know about the creative industry, who plays a key role in bottom–up regeneration. UK government the Department of Culture, Media and Sport (DCMS) defines creative industry as "those industries which have their origin in individual creativity, skill and talent and which have a potential for wealth and job creation through the generation and exploitation of intellectual property". DCMS also recognized 9 sectors, namely: 1) Advertising and marketing, 2) Architecture, 3) Crafts, 4) Design: product, graphic and fashion design, 5) film, TV, video, radio and photography, 6) IT, software and computer services, 7) Publishing, 8) Museums, galleries and libraries and 9) Music, performing and visual arts (DCMS, 2001). Some scholars also add toys & games, the R&D in science and technology (Howkins, 2001) as well as education industry (Hesmondhagh, 2002). Richard Florida argues that, human's creativity is the ultimate economic resource (Florida, 2002). Some other scholars state that the industries of the twenty–first century will depend increasingly on the generation of knowledge through creativity and innovations (Landry & Bianchini, 1995). Creative industries have been seen to become increasingly important to economic well–being.

These are a specific group of people. Due to their high dependence on the ability to generate creativity, they have similar characteristics and demands. Creative Communities are places where they work, live and have their leisure. Afterward, it is they who become the key players in the bottom–up regeneration.

What is their relation to space? What are the characteristics of space they demand? The answer to these questions can explain why, ever since they enter the existing city areas, creative industries continuously reconfigure the space that are related to them, and thus trigger the bottom–up regeneration. Such reconfiguration ranges from interior renovations of buildings, reconstruction or expansion of the buildings to the joint efforts to enhance the public space. Such comprehensive spatial reconfiguration is hardly found in any other type of modern office area. It leads to an assumption, which is that, the reason for spatial reconfiguration has a strong association with the characteristics and inner demand of creative industries and the related people.

One of the distinguished phenomenon is that numerous creative enterprises have sustained and great enthusiasm for modest working space. In a world which has gone global, with the refining segmentation of industrial network, the scale of many enterprises are getting smaller. Quite a few small–scaled ventures occupy specific and irreplaceable location in the globalized industrial network by retaining its creativity which grants them exceptional competitiveness and vitality, without having to expand their size to cover the complete industrial chain. In plenty of countries, small and micro enterprises (SME) [iii] have an overwhelming advantage to large enterprises in terms of the quantity of enterprises as well as the total number of employees. They are contributing a considerable section of their countries' economic output. Although the definition of SME and creative

图 14 M50 里的一处办公空间
Fig. 14 A working space
in M50 of Shanghai

办公空间本身也可以是公共空间。对于很多创意企业来说，与不同的人员进行交流，就是他们工作的一部分。因此，他们的工作空间天然具有极强的公共属性。

Office space itself can be turned into public space. For many creative enterprises, to communicate with different people are part of their work. Therefore their working space naturally has strong public features.

图 15 上海八号桥里的
公共展示空间

Fig. 15 The shared event space
in Bridge 8 of Shanghai

展览和酒会是这个空间里最频繁发生的活动。在没有活动的时候，这里也可能成为周边企业办公人员驻足聊天的场所。

Exhibitions and parties are most frequent activities in this space. When there are no formal activities, this space can turn into a place where employees of the neighboring enterprises pause and chat.

enterprises follows different logic, it is undeniable that the majority of creative enterprises are SME. They prefer affordable pint-sized working places and have high demands on the diversity of form in the spaces. Also, they will prefer changeable spaces adapted to their growth, which is dynamic according to enterprises' performance and development. For these reasons, urban village or Li Long housings spontaneously are turned into their ideal options. Another case is 798 Art Zone, in spite of its massive space units, it allows a flexibility of being divided into petite cells freely.

On the other hand, working spaces are places of exhibiting their products, likewise they may shift to parts of the brand image in different situations. For some enterprises, working spaces themselves are a component of their product. This consequently motivates them to select identified urban area with special spaces. Additionally, more than half enterprises show remarkable efforts and creativity to the renovation of the building's interior and exterior space. In Bridge 8 and M50 of Shanghai, we see a tremendous diversity in the interior space refurbishment of working space. Large numbers of working spaces accomplish their outer walls with glass disclosing the inner space and working scenes in order to welcome passing tourists or other creative people (Fig. 11,12,13). In Tian Zi Fang of Shanghai, the characteristics of historical Li Long housing become an influential element of the enterprises' brands, while the special space reconfiguration built by the enterprises contributes new virtues that differentiate Tian Zi Fang from the other Li Long housing areas.

The most exciting and admirable phenomenon in most creative communities is that, after their individual actions on their own working spaces, those creative enterprises eventually will act jointly on public space, which lead to the structural regeneration of the urban area as a whole. To produce new knowledge is the key function of creative industries, as well as a key feature that differentiates creative industries from other types of industries. New knowledge is produced basically from the association and cooperation between creative units (enterprise or personnel), which can only take place in public space physically. The existence of public space makes the essential condition of being a "community". From such perspective, we can answer why most bottom-up reconfigurations of space, initiated by individual industrial enterprises, evolve eventually from "individual action" to "joint action", whose main battlefield is on public space. The shared public exhibition area in Bridge 8, the alleys in Tian Zi Fang and the outdoor space for art activities in Sha Po Wei, are all consequences of such "joint action"(Fig. 14,15,16).

图 16 田子坊里的街巷空间
Fig. 16 Alleys in Tian Zi Fang

这些街巷原来是里弄内部的交通通道，现在变成游客聚集的街道。街道之间相互连接成一张密集的网，而每个住宅单元（现在被改造成各种小店铺和艺术空间），都尽可能对着这张路网来打开。

These linear spaces, which use to be semi–private lanes in Li Long housing communities, evolve into urban streets in which many tourists stay. They are further connected to each other until an entire net is formed. Most housing units, which are now transformed into shops and art spaces, are so much willing to be accessible from this net.

图 17 沙坡尾里的户外艺术空间
Fig. 17 Outdoor art area in Sha Po Wei

在各种类型的建筑之间，空地都变成了艺术活动和商业活动发生的地方。和城市里那些为了公共活动而被特意规划的街道和广场不同，这里的公共空间有着形式上和内容上不可比拟的多样性。

Between various typologies of buildings, those void spaces become place where art and commercial activities take place. In terms of forms and programs, they have a bigger diversity compared to those streets and plazas which are planned specifically for public activities.

3 混杂背后的秩序，
当代产业和城市空间的自组织

ORDER BEHIND CHAOS,
SELF–ORGANIZATION OF CONTEMPORARY INDUSTRIES
AND URBAN SPACES

3.1 复杂系统科学中自组织理论
SELF–ORGANIZATION THEORY IN THE SCIENCE OF COMPLEX
SYSTEM

3.2 当代产业作为自组织的复杂系统
CONTEMPORARY INDUSTRIES AS SELF–ORGANIZED COMPLEX
SYSTEM

3.3 城市空间的自组织特征及其面临的挑战
SELF–ORGANIZATION IN URBAN SPACE AND THREATENS IT
FACES

创意社区的空间形态是非常复杂和动态的，也许"混杂"是我们对该空间形态特征最贴切的描述。这与当代城市规划追求的"清晰的"、"可读的"结构是迥异其趣的。

是不是"混杂"就代表着"无序"？星辰（图18）、雁阵（图19）、蚁巢（图20）都可能显得非常无序，但我们知道，它们的形态其实都是由隐含的秩序控制着的。欧洲传统城市复杂的空间系统，几百年来从未停止过动态的变化，它们成功地容纳着无比复杂的功能、人群和多姿多彩的城市活动。这些城市的空间形态的确是混杂的，然而它们所呈现的那种令人叹为观止的多样性、弹性和给予我们视觉感官的巨大愉悦，却是那些拥有着"清晰结构"的当代新城区们所望尘莫及的（图21）。难道，这样的城区是"无序"的吗？

显然不是的。它们的秩序来自那些复杂的功能和人群之间千百年来不断地博弈所形成的土地和建筑的微妙关系。它如同人的神经系统般交织缠绕着，给予城市的肢体无处不在的支撑。秩序的复杂性和动态性则决定了作为其表征的空间形态的复杂性和动态性。

那么，对于创意城区而言，是不是也存在这样一种隐含的秩序决定着创意社区的空间形态？我们的回答是肯定的。包含原住民、租用原始房屋的创意企业以及城区管理部门和那些小的开发商的"利益相关者"之间的千丝万缕的关系，在没有或较少外力的干涉下，将转化为一种发展力量，推动着物质环境的升级、优化甚至重构，这就是"自发性城市更新"的内涵。

Chaos, this may be the strongest impression when we commence analyzing the complex and dynamic spatial form of creative communities. It is distinct from the "ideal city form" characterized by clear and readable structure, which is pursued by modern urban planning.

Is chaos disordered? Galaxy (Fig. 18), wild goose array (Fig. 19) and nest of ants (Fig. 20), they all appear disordered. Conversely, science tells us, their shapes come into being through a mechanism in which a complex hidden order is in behind. The highly comprehensive spatial system of traditional European cities, which has been a consequence of unceasing changes, successfully incorporates varied functions, people and various urban activities. That is to say, it is exactly the complexity of spatial form that provides an unusual flexibility and diversity, and aesthetically pleases to human senses. In contrast, these values are hard to be delivered by those contemporary cities equipped with "clearer structure"(Fig. 21). The question is, are these traditional cities disordered ?

Certainly not. The order for these cities derives from the subtle relationship between lands and building, formed through functions' and people's interaction during the thousand years. This order acts like human nerves that interlaces everywhere to give the city's body comprehensive support. And the complexity of order and its dynamic change lead to complexity and dynamic change of spatial form of the city.

Is there such hidden order in creative communities, which determines eventually their spatial forms? Our answer is positive. Without strong external intervention, the interlaced relation between stakeholders (original inhabitants, creative enterprises who rent their houses, administrators of the area and those small developers), will become a power that pushes the reconfiguring or even the restructuring of the physical environment. This is the connotation of bottom–up urban regeneration.

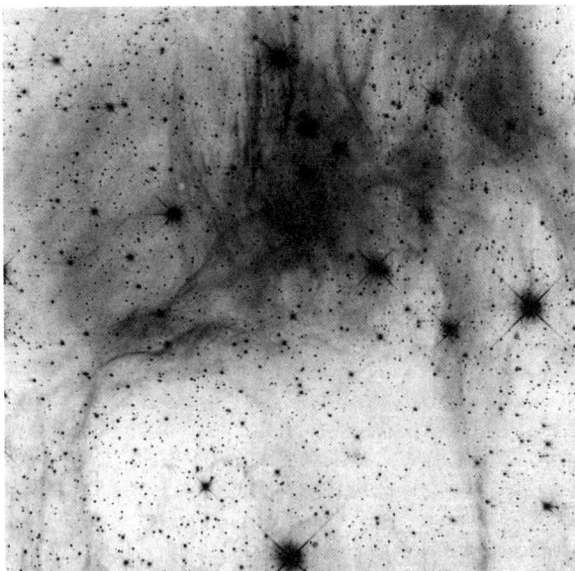

图 18 星辰
Fig. 18 Galaxy

图 19 雁阵
Fig. 19 Wild Goose Array

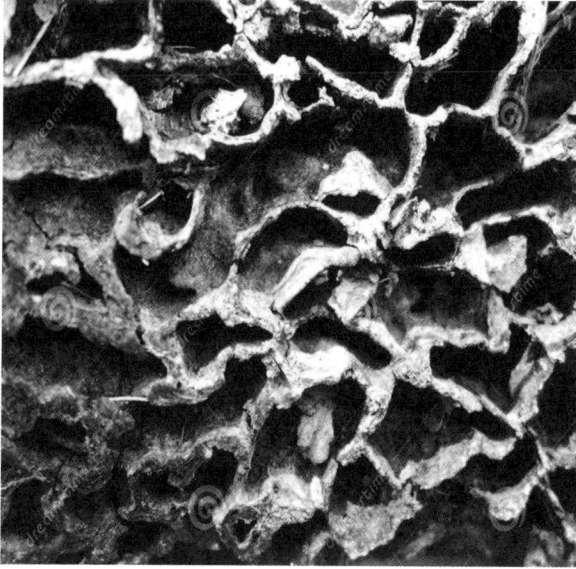

图 20 蚁巢
Fig. 20 Nest of ants

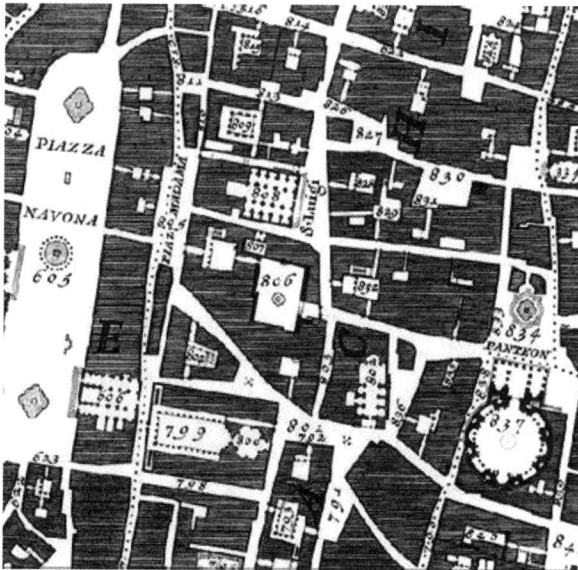

图 21 诺利地图的局部（罗马）
Fig. 21 Part of Nolli Map on Roma

3.1　复杂系统科学中的自组织理论
SELF–ORGANIZATION THEORY
IN THE SCIENCE OF COMPLEX SYSTEM

自组织理论是 20 世纪下半叶兴起的一种关于复杂系统的理论，它的基础是普利高津 (Prigogine) 的耗散结构理论、哈肯 (Haken) 的协同学、托姆 (Thom) 的突变论、曼德布罗特 (Mandelbrot) 的分形理论和洛伦茨 (Lorenz) 的混沌理论。它的主要观点是，一个远离平衡的开放系统，在保持和外界持续的能量输入以及内部子系统和构成要素的非线性作用下，系统不断地结构化、层次化，自发地由无序走向有序，或由有序走向更有序（图 22）。

自组织理论，是一个"全维度"的理论，它同时关注宏观（系统）、中观（子系统）和微观（要素）多个层面，并研究各个层面的相互关系。最早提出自组织相关理念的亨利·贝纳德，在他发现水分子规律运动和整体运动关系的时候，已经揭示了个体在系统里的重大作用（图 23）。在开放的复杂系统里，个体和系统必然是相互依存的，而整体的发展也必然与个体的运动方式有直接联系。

自组织理论，还是一个有深刻哲学启示的理论。第二热力学认为，对于封闭系统而言，"熵增"是不可避免的。城市、人类社会或者宇宙，如果它们都是封闭的系统，都必然遵循相同的热力学规律。基于此，第二热力学提出了宇宙必将走向"热寂"的假设（即极高的熵，极低的 秩序），这多少是悲观的和宿命论的。而自组织理论的提出，让这个假设变得不那么可怕了，因为如果这些系统不是封闭的而是开放的，那么维持系统的开放性并不断摄入能量，让系统中的众多个体在这个条件下充分互动，系统会向更复杂和更有序的方向发展。

自组织理论应用在自然和社会科学研究方面的巨大成功，不仅是在工具论层面（它建立了我们研究复杂系统的全新方法），更是在认识论层面上的，它让我们以全新的视角来认识这个世界和它的组织规律。

图 22 火车轨道的转接区
Fig. 22 Shifting area of rails

Having emerged since the second half of the twentieth century, Self-organization theory is a theory about complex system. It is based on the theory of dissipative structure, the theory of synergy, the theory of catastrophe and theory of chaos. Its main point of view is that an open system far from equilibrium, with the continuous energy input from the outside and the nonlinear action of the internal subsystems and its constituent elements, is continuously structured and layered, and spontaneously moves from disorderly to orderly state, or from orderly to more orderly state (Fig. 22).

Self-organization theory is a full-dimensional theory, which focuses on macro-, meso- and micro-level at the same time and aims to reveal interrelation between levels. Henri Benard, One of the earliest scholar who has great contribution to self-organization theory, has revealed the importance of individuals in the system, in his observation in relation between regular movement of water molecules and movement of water as a whole system. In an open and complex system, individuals and the system must be interdependent (Fig. 23). The development of the system is directly related to the way an individual moves.

Self-organization theory is also a theory with profound philosophical enlightenment. The second law of thermodynamics states that "entropy increase" is inevitable in an isolate system. A city, human society or the universe, if they are isolated system, will follow this path. Based on it, such hypothesis is put forward, which is the universe will eventually move towards "heat death" (that is, the ultimate high entropy and the ultimate low degree of order), which is somewhat pessimistic and fatalistic. The self-organization theory makes this hypothesis less scary, and the way to prevent heat death is to make the system open and constantly ingest energy, so that many every individual in the system can fully interact with each other under this condition. Simultaneously, the system moves in a more complex and orderly direction.

The success of the application of self-organization theory in natural and social science research is not only at the level of instrumentalism (which establishes a new method for us to study complex systems), but also at the level of epistemology. It gives us a whole new perspective on the world and how it is organized.

图 23 贝纳德对流花纹
Fig. 20 Benard Cells

3.2 当代产业作为自组织的复杂系统
CONTEMPORARY INDUSTRIES
AS SELF-ORGANIZED COMPLEX SYSTEM

自组织理论已经影响了自然与社会科学研究的各个方面。在产业研究方面，从韦伯的"产业区位"(A·Weber, 1929) 中对地理区位对产业集聚的研究，克鲁格曼的新经济地理模型 (P·Krugman, 1991)，到制度经济学提出的"边际交易成本"（R·H·Coases, 1960, 2014）等概念，显示对产业集聚的认识慢慢从关注企业个体行为转向关注产业集群作为一个动态系统。越来越多人认识到，当代产业是一个自组织的庞大系统。有的学者用"产业生态"来描述当代产业系统，认为基于进化论的传统生物学主要关注生物个体，在解释一些群体现象时是有局限性的，应该引入自组织理论的基本原理和方法才能够对产业系统的自我形成、自我调节、自我演化和自我消亡进行解释（孟薇、钱省三，2006）。产业系统的自组织行为，是基于集群内各个经济主体追求自我发展并为此展开的相互作用而发生的，且这些行为是在"特定的历史及环境条件下"自发产生的（王玲，2005）。

以上述理论为基础进行推断，创意社区中的创意产业，是以一种产业生态系统的形式存在的。该系统的发展，是基于产业个体的相互作用，因而系统的发展必然遵循自组织的相关规律。得出以上结论，是基于如下三个观察：第一，构成创意社区里的产业单体规模都比较小，灵活度高，生长性好，产业之间关联和互动非常密切。第二，创意社区的生长过程较少受到外界的指令影响（即指有组织的"规划"），一般是通过各利益相关者（Stakeholders）的相互作用来推动。相互作用的原动力是个体追求自我发展和共同发展。第三，构成创意社区的产业生态系统还不断地吸纳其他的产业类型以形成更大的系统，也和城市的其他功能发生强烈的联系（如城市中的居住和服务业功能等），而这种"开放性"正是复杂系统特有的。

Self-organization theory has influenced various aspects of contemporary scientific research. In the field of industrial research, the progressive theories from Weber's study of the geographical location of industrial agglomeration (A Weber, 1929), Krugman's new economic geography model (P Krugman, 1991), to the "marginal transaction cost" (R H Coases, 1960, 2014) show that the understanding of industrial agglomeration has gradually changed from individual behavior of enterprises to industrial clusters as a dynamic system. Some scholars use the term "industrial ecosystem" to describe contemporary industries, and point out that the traditional biology based on the theory of evolution is mainly concerned with the biological individual, which is limited in explaining some group phenomena. Thus the basic principles and methods of self-organization theory should be introduced to explain the self-formation, self-regulation, self-evolution and self-extinction of industrial systems (Meng Wei, Qian Huisan, 2006). The self-organized behavior of industrial system is based on the interaction between the economic individuals (enterprises) in a cluster and their pursuit of self-development. These behaviors occur spontaneously "under certain historical and environmental conditions" (Wang Ling, 2005).

Based on the above research, it can be inferred that the creative industry in the creative community exists in the form of an industrial ecosystem. The development of the system depends on the interaction of industrial individuals (enterprises). Therefore the development of this system must follow the law of self-organization. This argument is based on the following three observations: first, the creative enterprises in creative communities are relatively small-scaled with high flexibility and good growth. Their correlation and interaction is close. Second, the growth process of creative communities is less influenced by outside instructions, such as top-down planning, rather through the interaction of stakeholders. The motive force of the interaction is the enterprises' pursuit of self-development and common development. Third, the industrial ecosystem is also constantly absorbing other industrial types to constitute a larger system, and is also strongly linked to other functions of the city, such as residential and service functions, etc. While being "open" is a unique characteristic of complex systems.

3.3　城市空间的自组织特征及其面临的挑战
SELF-ORGANIZATION IN URBAN SPACE
AND THREATENS IT FACES

　　城市，正是人类所能创造的最复杂的系统之一。如果我们认可城市空间必然反映其背后无数的利益相关者之间的关系这样的观点，那么我们也必然认识到，自组织本来就是城市的天性之一。

　　当我们翻开欧洲城市、伊斯兰城市或者东方城市历史街区的地图，那些蜿蜒曲折的街道、形式复杂而精巧广场和丰富多变的建筑物轮廓每每让我们心生赞叹（图 24）。这既不是造物主为之，也并非一蹴而就。一切都出自人类之手，在漫长的历史过程中慢慢雕琢和完善。这里的"人类"，并非是抽象的整体，而是一个个具体的人（当然，也包括人与人之间组成的大大小小的利益同盟），他们生活在这里，对自己周遭的环境有自己的认识和要求，也通过改变他们的周遭环境来彰显他们的存在。这些改变大部分情况下通过人们之间的协商或合作来实现，另一些情况下则演变成争夺甚至是战争。空间的改变迭代相传，才形成了今天我们看到的城市空间。

北京
Beijing

迪拜
Dubai

京都
Kyoto

罗马
Roma

威尼斯
Venice

维也纳
Vienna

图 24 一些欧洲、伊斯兰和亚洲城市的肌理
Fig. 24 Urban Fabric of Some Europe , Islamic and Asian Cities

上面几个城市的城市肌理反映了城市中建筑物和空间的复杂性，正是通过千百年来城市的自组织发展而形成的。当然，在这些肌理中我们也能看一些强有力的"规划"介入的痕迹，比如北京的路网，或者欧洲城市里的一些大型广场。

These urban fabrics reflect the complexity of buildings and spaces, which comes into being through city's self-organization in thousands years. Certainly, in segments in these fabrics, we also see traces of some powerful "plannings", such as the road net of Beijing and those big plazas in European and Islamic cities.

今天的状况是，自组织带来的城市空间的极大丰富性，正遭遇巨大的损害。当我们一次次地对当代新建城区里整齐划一的道路和建筑感到无比失望时，我们不得不进行反思，也许，造成这种状况的原因，并不是我们的规划不够精巧和复杂，而是我们根本就误解了城市的本质。工业化初期的知识分子对于理性规划有着巨大的自信，他们一度认为空间是机器，它也可以像机器一样被生产，用以满足大部分人的居住和其他需求。我们暂且不去纠结于从政治学和社会学对这种"空间生产"的批判（事实上，列斐伏尔和森内特已经在这方面做了大量工作[IV]），仅从城市作为复杂系统的这个事实入手，我们就可以提出很多质疑。例如，"自上而下"的空间安排是否就能代表所有利益相关者的空间需求？有差异性的需求又如何体现？又例如，即使这样的安排足够复杂和精巧，它又如何保证在城市发展的漫长周期里，空间能够即时反映利益相关者的变化？事实上，如果城市的空间能够被一些"自上而下"的范式完美呈现，那它作为复杂系统的一切都已经消失了，它只不过是一幅精美的图画，即便手握画笔的可能是令人赞颂的"上帝之手"。

我们必须承认，当代城市面临的一些问题（尤其是涉及全体利益相关者的问题，例如基础设施）只能通过一定的"自上而下"的手段来完成。但是，是否这样的空间安排必须渗入到城市空间的方方面面？如果不是，那么城市空间的哪些方面，是可以允许一定的弹性，让利益相关者参与其中的？抑或是说，"自上而下"和"自下而上"二者的边界在哪里？ 有学者提出将上述的提问概括为"规划不可规划之事"，这样的概括是非常有洞见的（波图加尔，2012）。

创意社区的出现，为我们研究上述问题提供了极好的实证案例。废弃的厂房变成艺术空间、历史住宅区被开放为商业空间、城中村中生长出来的多样化产业功能，这些创意产业推动下的看似"非法"的功能置换和建设行为，催生出如此精彩的建筑改造和空间改造，甚至完全改变了一个城区的空间结构。这些案例向我们证明，好的城区不一定是"规划"出来的，也可以是"生长"出来的。在这里，一种古老的、我们久违了的城市发展方式，复活了。它们可以给我们什么启示？

The city is one of the most complex systems that man creates. If we accept that the urban space necessarily reflects the relationship between the numerous stakeholders behind it, we must realize that self-organization is one of the natures of the city.

When we turn to a map of the historic neighborhoods of European cities, Islamic cities, or Asian cities, we marvel at the winding streets and intricate squares, as well as the rich and varied contours of buildings blocks (Fig. 24). This was not done by the Creator, nor was it done overnight. Everything was by the human hand in the long process of history slowly carved and improved. The "human" here is not an abstract whole, but one and another person (including, of course, the alliance of interests, big or small, among people) who were living here, who had their own understanding and demand on their own living environment. They changed their surrounding environment to highlight their existence. In most cases, consultation and cooperation assisted in achieving these changes, nonetheless, in other cases, they generated rivalries or even evolved wars. Space changes passed down iteratively, and the urban space we see today was finally formed.

What is happening today is that the abundance of urban space brought about by self-organization is suffering enormous damage. We are disappointed again and again by the monotony of the roads and buildings in the new urban areas of our time, hence we have to introspect. Perhaps, it is not because our planning is not sophisticated and complex enough. Simply, we have misunderstood the nature of the city. In the early days of Industrialization, the intellectuals had strong confidence in rational planning and considered space a machine which could be manufactured to satisfy the living and other needs of most people. Let us leave aside the political and social critique of this "production of space" (in fact, Lefebvre and Sennett have done a great deal of works on this [iv]), and start with the fact that the city is a complex system. Here we can raise a lot of questions. For example, does a "top–down" spatial arrangement represent the spatial needs of all stakeholders? What is the representation of differentiated needs? And even if such an arrangement is sufficiently complex and sophisticated, how does it ensure that the space can instantly reflect changes of demands of stakeholders during the long cycle of urban development? In fact, if urban space can be perfectly regulated by some "top–down" paradigm, it will lose all quality of a complex system, it is only a fine picture, even though that hand holding the brush is the praiseworthy "hand of God."

We must recognize that some of the challenges that contemporary cities face, particularly those involving all stakeholders, such as urban infrastructure, that can only be accomplished through a certain "top–down" approach. But does this spatial arrangement have to permeate all aspects of urban space? If not, what aspects of urban space may award a certain degree of flexibility, allowing stakeholders to participatin its creation? Where is the interface botwccn "top–dowii" approach anWd "bottom–up" growth? Some scholars define these questions as "to plan the unplannable", it is an insightful recap (Portugali, 2012).

The emergence of creative communities provides us with excellent empirical cases to study the above issues. Abandoned workshops become art spaces, historical residential areas are open to commercial activities, and diversified industrial functions grow out of villages in cities. These creative industries promote seemingly "illegal" functional replacement and construction, give birth to a wonderful architectural and space transformation, which in the end leads to complete changes in the spatial structure of an urban area. These cases have proven that superb urban areas are not necessarily "planning" out, they can also be "growing" out. Here, an ancient, long–lost way of urban development, revives. What lessons can we learn from them?

4 创意社区自发更新的动力和空间表征
DYNAMIC MECHANISM AND SPATIAL CHARACTERISTICS OF THE BOTTOM-UP URBAN REGENERATION OF CREATIVE COMMUNITIES

4.1 创意产业为什么集聚？
WHY DO CREATIVE INDUSTRIES AGGLOMERATE?

4.2 利益相关者如何影响空间？
HOW DO STAKEHOLDERS AFFECT SPACE?

4.3 空间的动态性特征
DYNAMIC CHARACTERISTICS OF SPACE

4.4 空间与产业的关联
ASSOCIATION BETWEEN SPACE AND INDUSTRIES

4.5 空间的类型化特征
TYPOLOGIES OF SPACE

这个章节将展示我们实证研究的主要发现。我们挑选了五个中国的案例：上海的田子坊、厦门的沙坡尾和曾厝垵、北京的 798 艺术区以及深圳的大芬村。我们将研究的发现分为"内因"和"表征"两个部分来呈现，第一部分讨论关于自发更新的"内因"，即，创意产业为什么聚集，又是什么在推动自发更新以及该推动力如何通过一定的机制作用于空间。第二部分则讨论"表征"，即创意社区的空间形态作为一个物质结构的特征，以及这些特征如何反映其背后的推动力及其作用机制。

4.1　创意产业为什么集聚？
WHY DO CREATIVE INDUSTRIES AGGLOMERATE?

创意产业在城市中的集聚，既源自企业的经济行为，也是一种社会行为。当代城市产业选址的总体规律是"去中心化"（decentralization），即制造业远离城市中心，转移到远郊或者在全球化环境中重新选址（许凯，克劳斯·泽姆斯罗特，2013）。但是，包括创意产业在内的都市产业（Urban Industry）有较高的地租支付能力，并与其他城市功能关系密切，它们倾向于选址在城市中心区域。这可以运用微观经济学的成本 / 盈利曲线（D·M·史密斯，1971）或者是行为学派（Behaviorism）的选址理论（D·基布尔,1976）加以解释（该理论提出应纳入"生活环境"、"劳动力市场供求"和"城市政策"等其他选址要素解释产业的选址行为）。

但是，同样在城市中心区，创意产业会具体选择什么类型的地块入驻，是一个需要进一步研究的问题。笔者认为，在上述选址因素之外，应该纳入"空间因素"。创意产业选址决策者是创意企业和企业的从业人员（即"创意阶层"ˇ），企业和人员认可什么样的空间特点，是选址的重要原因。佛罗里达在《创意阶层的崛起》中强调的"场地特质"包括：创意环境、互相合作的社区成员、文化艺术活动、街道活动及咖啡馆中的聚会等创意事件（佛罗里达，2014）。佛罗里达的这段描述并没有特指空间属性，但隐含空间的某些特征，即空间需要满足多样性的人群和活动的发生。此外，同样非常重要的是所选城区的空间应具有容纳系统自组织发展的基础，即它可以在利益相关者的推动下发生灵活多变的空间调整，以满足系统的动态性发展。这可以帮助我们解释为什么创意社区选择一些城区（如废弃的工厂、城市历史居民区和城中村等），而不选择一切其他的一些城区（如配套更好的办公园区、商业区和住宅区）。

The main findings of our empirical research are to be presented in this chapter. We elected five Chinese cases: Tian Zi Fang in Shanghai, Sha Po Wei and Zeng Cuo An village in Xiamen, 798 Art Zone in Beijing and Da Fen Village in Shenzhen. We present collected data in two parts: : "internal cause" and "spatial representation". The discussion on the "internal cause" of bottom-up renewal opens in the first part as following, why creative industries agglomerate, what is the driving force for spontaneous renewal and how the driving foxrce acts upon space through a certain mechanism. The second part discusses "spatial representation", in brief, as a material structure, how the characteristics of spatial forms of the creative communities reflect the driving force and mechanism behind those spatial forms.

对案例的研究主要关注三个方面（表1）：

第一是城市本身的特点，及其和案例中的产业类型的对应关系。比如深圳是国际贸易和物流发达的城市，油画主要面对的是国际的市场，因此大芬村具有吸引油画产业入驻的基本条件。厦门的旅游城市特质则催生了曾厝垵的精品酒店业的发展。北京和上海作为中国文化艺术中心城市，为798艺术区之类的艺术产业提供了基础。

第二是案例在城市中的区位。大部分案例都坐落于城市发展核心区域邻近的城区，这让创意产业从业人员可以较容易从周边城区获得城市服务，和比较低的通勤成本。此外，选址周边的一些资源如大专院校或旅游目的地，对选址有比较好的支持作用。例如曾厝垵和沙坡尾这两个案例都邻近厦门大学，且周边旅游业发达，学生、教师和游客都对产业发展有一定支持。

第三是城区本身的空间环境是否存在建筑的或者空间的"特色"，这是创意企业和人员选址的重大因素。例如"历史城区"（沙坡尾、田子坊）、"城中村"（大芬村、曾厝垵）和"特色工业区"（798艺术区）这些"特色"都为相应的创意社区提供独特的文化品牌。此外，实地调研也发现城区中的特定空间类型为产业活动提供了支持。大芬村的主要街道成为油画企业生产的场地和对外招徕业务的空间，沙坡尾的滨水空间和曾厝垵的主要街巷是特色商业的聚集地，而798艺术区里的大跨空间厂房则成为举办展览的最佳场所。

The agglomeration of creative industries in cities is not only an economic behavior of enterprises, but also a social behavior. The general inclination of industrial location selection in contemporary cities is "decentralization", that is, the manufacturing industry moves away from the city center to the outer suburbs or relocates in the locations in global city networks (K XU, K Semsroth, 2013). However, urban industries, to whom the creative industry belongs, have a higher capacity to pay land rent and has a close relationship with other urban functions. Therefore, they tend to be based in city's central areas. The cost / profit curve of microeconomics (D M Smith, 1971) can elucidate this tendency, so can the theory of location in Behaviorism (D Keeble,1976), which suggests that "living environment", "supply and demand of labor market" and "urban policy" should be taken into account to explain the behavior of industrial location.

In the city's central area, what types of location are the preferable location for creative industries? We hold that, in addition to the above-mentioned economic and social factors of location selection, "spatial factor" shall also be included. In location selection of creative industries, the decision makers are those creative enterprises and related personnel, who naturally belong to "creative class"[v]. What kind of spatial characteristics these enterprises and personnel recognize is an important reason for their location selection. The "site parameters" highlighted by Florida in <The rise of the Creative Class> include creative environments, collaborative community members, cultural and arts events, street life, and gatherings in cafés (R Florida, 2014). This description does not specifically refer to certain spatial attributes, but implies some of the characteristics of space, that is, the need for space to meet the diverse people and the occurrence of activities. In addition, it is important that the selected urban space should have the flexibility to accommodate the self-organized development of the community, that is, it can be adapted spatially to reflect the needs of changing stakeholders. This helps explain why creative communities choose some urban areas (such as abandoned factories, historic urban neighborhoods and urban villages), instead of some other urban areas (such as better equipped office parks, business districts and residential areas).

The case study focuses on three aspects（Tab.1）:

The first is the characteristics of the city itself, and its relation to industry types in the cases. For example, Shenzhen is a highly developed city of international trade and logistics. Therefore, Da Fen Village has the basic conditions to attract the oil painting industry, which is mainly oriented to the global market. The characteristics of Xiamen as a tourist city hastened the development of Zeng Cuo An's boutique hotel industry. Beijing and Shanghai, as the center cities of Chinese culture and art, provide the foundation for the art industry of 798 Art Zone and Tian Zi Fang.

The second is the location of the case in the city. Most of the cases are located near to the city' s core urban development areas, which makes it easier for creative enterprises to get access to services from the surrounding areas, and the cost of commuting is also lower. In addition, some resources, such as universities or tourist places, can further support the location of creative communities. For example, Zeng Cuo An and Sha Po Wei are both close to Xiamen University, as well as many other tourist attractions. Students, teachers, and tourists all have certain supports for the development of the creative industries in both cases, because these people may become employers, employees or customers of the creative enterprises in both cases.

The third is whether the selected urban areas have strong architectural or spatial identity, which is a substantial factor in the location preference of creative enterprises and personnel. For example, "historic urban quarter" (Sha Po Mei）, "urban village" (Da Fen Village and Zeng Cuo An Village) and "industrial heritage" (798 Art Zone) all present unique cultural brands corresponding to creative communities. In addition, field studies have discovered that specific spatial types in urban areas support industrial activities. The main street of Da Fen Village has become the production site of oil painting enterprises and the space for attracting business to the outside. The waterfront space at the end of Sha Po Wei and the main street of Zeng Cuo An Village become the outstanding places of featured commerce. In the 798 Art Zone, the large–span factory space is the quintessential place for artists and art enterprises to hold exhibitions.

表 1 创意产业选址的空间要素
Tab. 1 Spatial factors for creative communities to select their locations

案例 Cases	区位特征			建筑和空间特征		
	城市特点 Identity of the city and how it supports the industries	与城市中心的距离 Distance from city's center	周边对产业发展有支持的资源 Resources in surroundings that may support the industries	作为"文化品牌"的空间或建筑特征 Space or architeure identities that become brand of the area	适合特定产业发展的建筑空间特点 Characteristics of buildings that fit the demand of new industries	适合产业发展的外部空间 Characterietics of outdoor space that fit the demand of new industries
沙坡尾 Sha Po Wei	厦门是著名的旅游城市，对文化特色和创意商业的需求，支持沙坡尾的产业发展 Xiamen is a famous tourist city. The demand for cultural identity and creative commercial business can support the the creative industries of Sha Po Wei.	距离城市中心（中山公园）2.1km 2.1 kilometers from the city center (Zhongshan Park)	厦门大学作为文化教育中心，为文化艺术产业提供知识和人力资源；此外，临近中山路历史街区和环岛路的地理位置，让沙坡尾获得充足的游客资源 Xiamen University, as a cultural and educational center, provides knowledge and manpower for the cultural industry. In addition, its location, being close to the historical zone of Zhongshan Road and main tourist attraction of Round-island Road, helps Sha Po Wei to get enough tourist flow.	渔港、连续的滨水岸线和富有历史特色的骑楼建筑群 Fishing ports, continuous waterfront shoreline and historic arcade buildings.	较小的可以切分出租的空间单元，适合小型产业核消费空间 Smaller space units that can be divided into smaller parts for rent. Buildings are suitable for small enterprises and shops.	滨水空间、滨水广场节点和骑楼能支持产业和商业的发展 Waterfront space, waterfront plaza nodes and arcade can support the development of industries and commercial business.
曾厝垵 Zeng Cuo An Village	游客对文化特色尤其是特色酒店的需求，支持曾厝垵的产业发展 The demand for boutique hotels with culture identity supports the development of Zeng Cuo An's featured industry.	距离城市中心（中山公园）4.6km 4.6 kilometers from the city center (Zhongshan Park)	厦门大学艺术学院支持曾厝垵早期的发展，提供一定创意人群和消费人群；临近环岛路旅游区是酒店业获得发展的重要原因 The school of art of Xiamen University supports the early development of Zeng Cuo An, providing creative people and consumers. Being near to the Round-island Road, it can get enough tourists for the development of the hotel industry.	闽南乡村特色的街巷和一些特色历史建筑 Streets, lanes and some distinctive historic buildings with Southern Fujian Province style.	小体量独栋建筑，可以完整或切分运营，特别适合精品酒店经营，同时也适合特色商业经营 Small-volumed single buildings can start an integrated or segmented operation, particularly suitable for Boutique hotel operations, and also suitable for special commercial operations.	村口广场和村中的主街成为酒店和商业聚集的主要地点 The village plaza and main street are the main locations for hotels and commercial businesses.
798艺术区 798 Art Zone	北京作为全国文化艺术中心城市，对文化艺术产业有较大的需求，使798艺术区具有辐射全国的产业影响力 As a national center for art and culture, Beijing has a great demand for cultural and art industries, making 798 Art Zone's nation-wide influence .	距离城市中心（天安门）11.9km，北京CBD 9.5km，望京CBD1.8km 11.9 kilometers away from the city center (Tiananmen), 9.5 km from Beijing CBD, and 1.8 km from Wangjing CBD.		现代主义风格的厂房 Modernist-styled workshop	大体量、独栋的大跨厂房，空间艺术特色鲜明强烈，适合做展览及艺术家工作室 Large-volumed and wide-spanned workshops have strong artistic space features, which is suitable for exhibitions and workshops for artists.	三处广场是主要的室外开放展览空间 The three plazas are the main outdoor exhibition spaces.
大芬村 Da Fen Village	深圳作为全球商业贸易和物流中心城市，对艺术产业的需求和支持 As a global trade center, Shenzhen demand and supports art industries.	距离深圳东站1.8km，深圳大剧院8.1km，深圳CBD11.5km 1.8 kilometers from ShenZhen East Station, 8.1 kilometers from the ShenZhen Grand Theatre, and 11.5 kilometers from the ShenZhen CBD.			小体量独栋建筑，适合画廊、仓储、油画生产和画工宿舍的功能使用，也同时可以保留一部分村民居住 Small-volumed villager houses can fit use of galleries, storage, painting pruduction and dommitory for painters. Parts of it can still remain residential use for villagers.	城中村的主街支持画廊对外招顾客的功能 The main streets in Da Fen Village are the place for galleries to attract customers.

38

4.2　利益相关者如何影响空间?
HOW DO STAKEHOLDERS AFFECT SPACE?

　　以创意企业和创意人员为基本单元的复杂系统，要通过一定动力机制，才能转化为空间，这与当代城市规划中"由图纸到空间"的转译方式是非常不同的。创意社区的空间形成与发展，是通过多样的"利益相关者"（stakeholder）的协商并通过"局部建设"来实现的。各利益相关者的竞争或协同，正是系统自组织的核心特征。创意社区中，各利益相关者的一般关系是怎么样的？它又是如何推动空间发展的？此外，在很多创意社区发展的后期阶段，当政府逐渐认可创意社区的存在，往往会制定一些自上而下的规划，力图进一步提升这些区域的品质。那么，这些自上而下的规划，作为一种"外部指令"，对系统的发展会产生什么影响？它在上述的动力机制里又占据什么地位？

　　首先我们来看看个体的作用，这始终是复杂系统发展的基本力量。对创意社区而言，最典型的个体就是租用房屋的创意企业和相关的创意人员。他们首先置换房屋的功能，进而对房屋按照他们的需求进行内部改造，就像我们在几乎所有创意社区中看到的那样。接下去，一些企业就会开始寻求更进一步的发展，这包括向相邻的单元进行扩张并将那些本来分散的空间进行连接。他们也当然会在可能情况下对建筑的立面和形体进行更改来适合企业的功能使用，例如大芬村和曾厝垵中大部分企业会加建楼层。很多的创意企业对公共空间会有一定的需求，用以增强与游客（很可能会成为企业的客户）的交流，这会驱使企业对建筑物的沿街面进行改造（"破墙开店"是其中最典型的），并改造和利用与自己房屋接壤的外部空间地面，如增加地面铺装和增设桌椅外摆。企业的数量增加并彼此相连时，大家就会发现将各自的外部空间整合起来规划是对大家都有利的，于是街道就产生了。沙坡尾的滨水街道就是这样形成的。

　　当然，实际情况可能更复杂一些。除了创意企业和创意人员，仍旧居住在区内的居民或区内既有功能的使用者（如在798案例中原厂房的所有者）也构成利益相关者的一部分。他们对空间当然有不同的需求。例如，有些居民可能就对喧闹而不那么安全的公共空间感到十分反感，而其他一些居民仍可能支持创意企业对外部空间的改造，因为这些居民作为出租房屋的业主在整个城区的发展中获益。居民和企业之间、居民和居民之间、企业和企业之间的那些联合行动或斗智斗勇，推动着创意社区的一次次的空间转变（图25、26、27）。

　　当我们尝试为上述的空间动力机制整理出一定规律时，我们发现，各个案例中利益相关者的构成和相互关系是非常类似的（图28）。原住民、租赁原住民住所各类产业从业人员构成了利益相关者的第一层级："个体"。这些利益相关者个体进一步组合成各种组织，如艺术家委员会、业主委员会、租户委员会、园区发展委员会等，构成利益相关者的第二层级："组织"。在这个层级之上，是园区管理委员会，作为协调利益相关者的组织，同时代表上级政府的意志。代表政府利益的国有开发公司和代表投资者利益的私人开发公司，则作为跨层级的利益相关者，它们既要代表各层级利益相关者的一定意志，同时还带有自身的利益诉求。

从利益相关者对空间形态的作用方式和能力上看，第一层级的利益相关者主要通过对"微观空间"的改造（如建筑物改造和与个体空间紧密相关的小型公共空间）来实现。这些作用时刻在发生，影响范围比较小，却是创意社区空间最基层的推动力。第二个层级的利益相关者则代表了创意社区一部分人群的共同意愿，这些意愿往往能够转化为对"中观空间"的干预，如区域主要的公共通道、广场和出入口等（通过合作的方式实现）。最后，在特定的发展阶段（往往在社区发展的后期），政府会以进一步发展社区为目的组织正式规划，形成对社区空间形态的整体性干预。这些规划作为对社区发展的"外部指令"，其执行过程可能会遭遇各利益相关者的抗拒。从其实际效果上看，也未必对社区的发展有利，例如审美异化、过度商业化（追求租金收益）和失去产业多样性等问题。从空间上看，则往往会倾向于抹去创意社区的复杂和动态的空间特征，转向较为清晰可读的结构，从而丧失原有空间的多样性。从积极的方面来说，部分规划的确能推动社区一定发展，如沙坡尾案例对滨水空间的改造、增厝垵案例对入口和主街的改造，都促进了社区的进一步发展，这种宏观层面上的结构性调整通过自组织方式很难实现的。在798艺术区案例中，正是因为缺乏宏观的调控导致空间发展陷入僵局[vi]。

原始空间
Original space

产业入驻并改变局部空间
Embeddedness of enterprises
and individual change in space

产业聚集并推动空间整体更新
Agglomeration of enterprises and
integral spatial regeneration

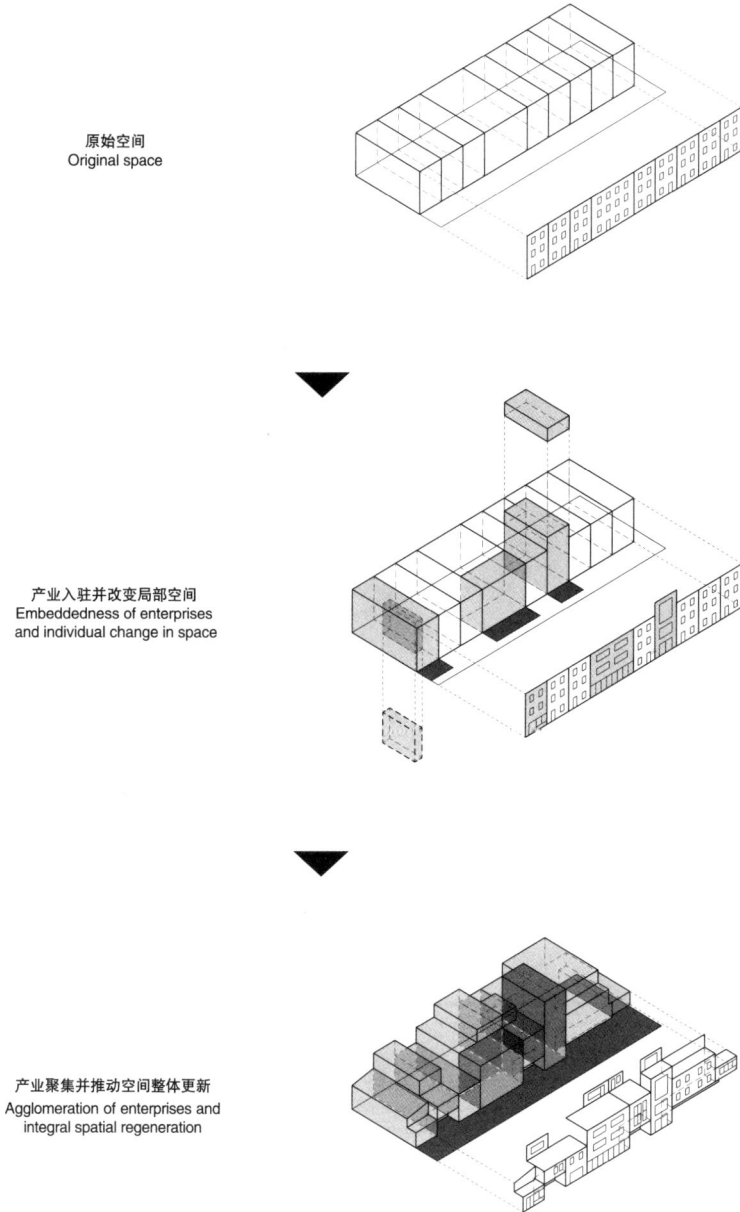

图 25 联排式空间单元和空间发展模式
Fig. 25 Row units and their development model

41

原始空间
Original space

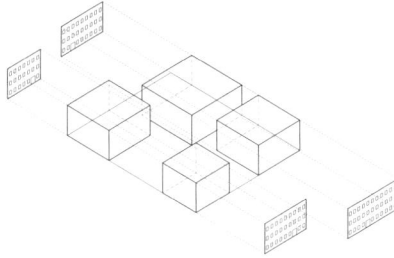

产业入驻并改变局部空间
Embeddedness of enterprises
and individual change in space

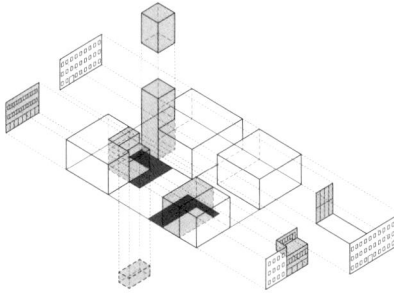

产业聚集并推动空间整体更新
Agglomeration of enterprises and
integral spatial regeneration

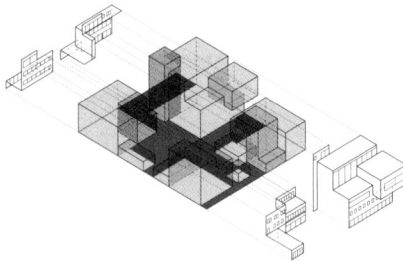

图 26 小型独立空间单元及其发展模式
Fig. 26 Small individual units and its development model

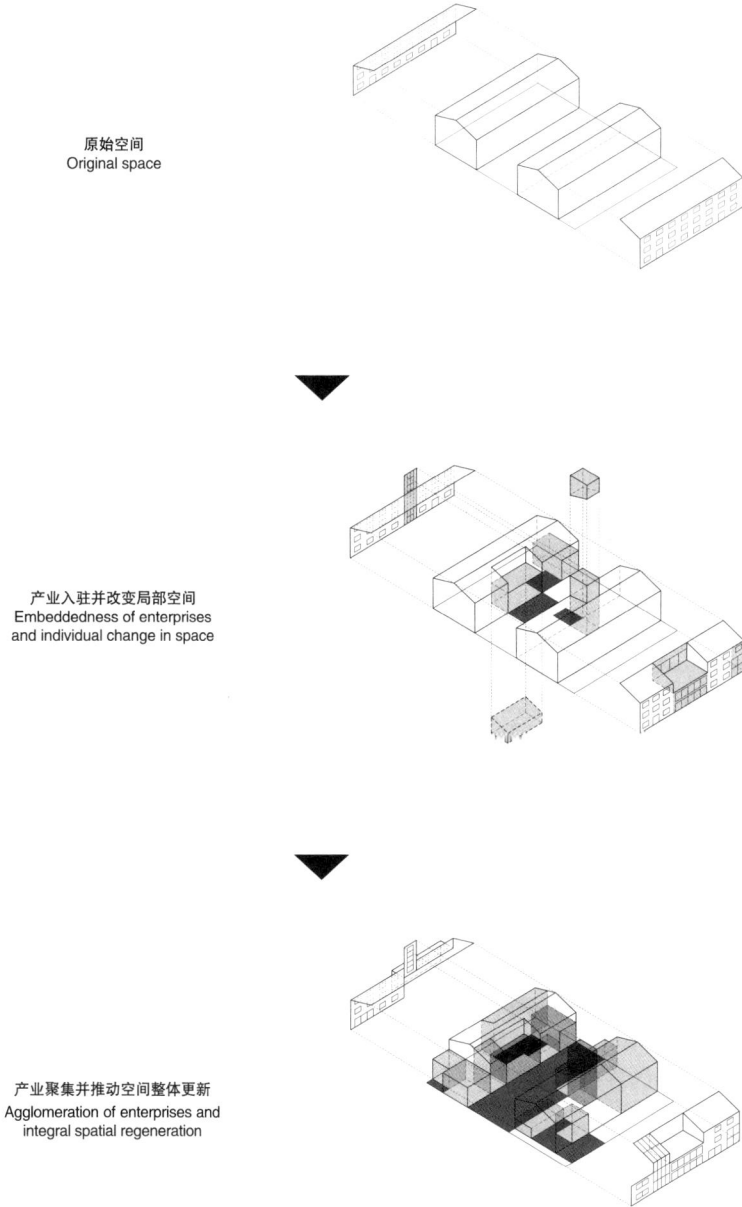

原始空间
Original space

产业入驻并改变局部空间
Embeddedness of enterprises
and individual change in space

产业聚集并推动空间整体更新
Agglomeration of enterprises and
integral spatial regeneration

图 27 大型独立空间单元及其发展模式
Fig. 27 Large individual units and its development model

The complex system, with creative enterprise and creative personnel as the individuals, can only be transformed into space through a certain dynamic mechanism, which is different to the "from–drawing–to–space mechanism" in contemporary urban planning. The spatial formation and development of the creative community is realized through the negotiation of various "stakeholders" and the "partial construction". Indeed, the competition or synergy among stakeholders is the core characteristics of the self-organizing complex system. What is the general relationship between the stakeholders in the creative community? How does it promote the spatial development? The research on mechanism also recognizes one situation, that is, in the later development stages many creative communities, the government gradually recognized the existence of the creative community and issue top–down planning attempting to further enhance the quality of these areas. What will be the impact of these top–down planning, as an "external directive", on the development of the system? What is its position in the dynamic mechanism mentioned above?

Let's look at the role of the individuals, which has always been a fundamental driving force in the self–organized complex systems. For the creative community, the most typical individual is the creative enterprises who rent the existing buildings and the related creative personnel. They first replace the function of the building, and change the inner space of the buildings to adapt to their needs, as we have seen in almost all creative communities. In turn, some businesses will start looking for further development, including expanding into adjacent units and connecting previously scattered spaces. They will, of course, change the facade and the shape of the building to suit the function of the enterprise. In the case of Da Fen Village and Zeng Cuo An, most enterprises will add floors to existing building to get more space. Many creative enterprises have a need for public space to enhance communication with visitors, who may well become corporate customer. This will drive enterprises to transform buildings along the street (the most typical of which is "breaking walls and opening shops"), and to transform and utilize the exterior ground space adjacent to their own buildings, such as adding floor pavement, plants and putting tables and chairs there. As the number of enterprises increases and are connected with each other, people will find it beneficial to integrate their external spaces as a common public space, thus the streets are created. This is how the waterfront streets at the Sha Po Wei have been formed.

Of course, the reality may be a little more complicated. Besides creative enterprises and creative personnel, staying residents or people who still use the remaining functions in the area (such as those factory owners in the case of 798 Art Zone), are also influential stakeholders. They certainly have different needs for space. For example, some residents may resent noisy, less safe public spaces, while others may boost creative enterprises' efforts to transform the outer space so as to benefit from the regeneration of the area by getting higher rent. It is an aligned action coupled with battles of wits and guts between residents. Rhythmically, that is propelling one spatial transformation of creative communities after another (Fig. 25, 26, 27) .

When we try to work out some general laws for the above spatial dynamic mechanism, we find that the composition and relationship of stakeholders in each case are similar (Fig. 28) . The first level of stakeholders is the "individuals", which is composed of original residents and creative personnel who work in various enterprises. These stakeholders are further grouped into various organizations, such as the artists' committee, the owners' committee, the tenant committee, the community development committee and so on, which constitute the second level of stakeholders, namely, "organization". Above this level is the community management committee, which acts as an organization for coordinating stakeholders and represents the will of the higher levels of government. The state-owned development companies serving the interestsof the government and the private development companies serving the interests of investors, as the cross-level stakeholders, not only represent they the will of the stakeholders at all levels, but also their own interest.

From the perspective of the role and ability of the stakeholders in the spatial form, the first level of stakeholders mainly acts on the small-scaled space (such as internal space of buildings and small public space closely related building units). These actions, whose influence scope is relatively limited, occur all the time. It is the most basic driving force of the creative communities' regeneration. Stakeholders at the second level represent the common will of a segment of the creative community, which can often be translated into intervening in the "meso space"; Such as the main public access, streets, squares and entrances, and so on (through the means of cooperation between stakeholders). Finally, at a specific stage of development, the government in most cases will issue top-down planning, which acts as an overall intervention in the spatial form of community. Being "external directives", the implementation of such planning may be resisted by various stakeholders. Judging from the actual effect in many cases, it may not be beneficial to the sustainable development of the community, as it often leads to aesthetic alienation, excessive commercialization (for the pursuit of rental income) and loss of industrial diversity. In terms of transformation of space, it tends to erase the complex and dynamic spatial characteristics, and turn to a more clear and readable structure, thus losing the diversity of the original spatial form in creative communities. However, on its positive side, some of the top-down planning can indeed promote the development of the community. For example, the transformation of waterfront space of Sha Po Wei, the transformation of the entrance and the main street in Zeng Cuo An Village, both would not have been realized without proper top-down planning, as such structural adjustment at the macroscopic level is difficult to achieve through self-organization. By comparison, in the case of 798 Art Zone, it is the missing of intervention in macroscopic structure that leads to the deadlock of its spatial development [vi].

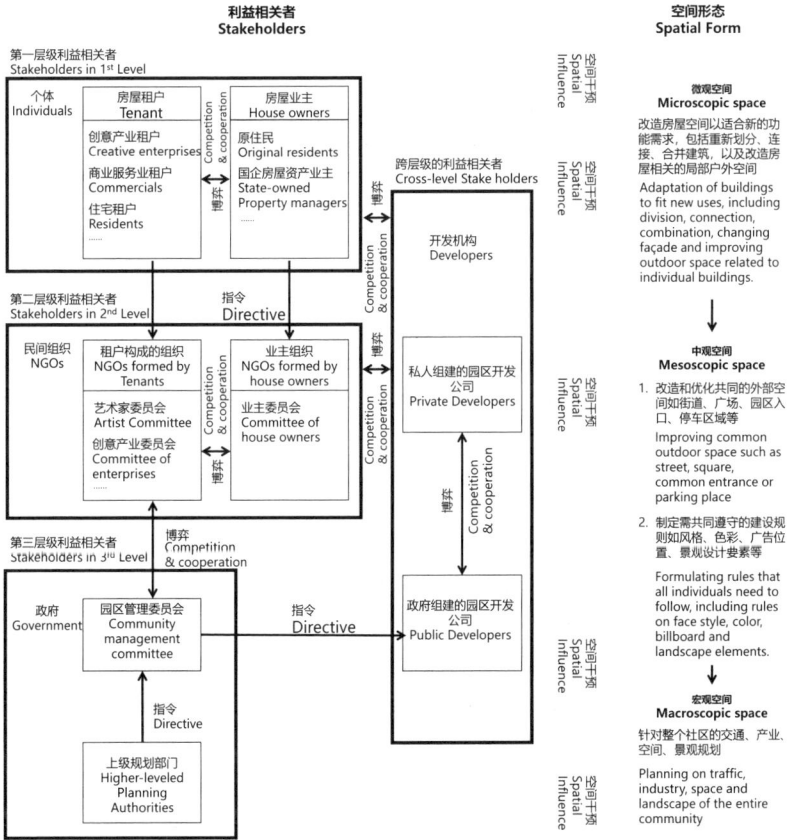

图 28 利益相关者谱系及其与空间形态的关系
Fig. 28 Pedigree of stakeholder and its influence on spatial form

4.3   空间的动态性特征
DYNAMIC CHARACTERISTICS OF SPACE

创意社区的发展是一个动态过程，它的空间结构必然是不断在变化的，这正是复杂系统发展的一个重要特征。根据自组织理论，复杂系统的发展是通过"涨落"[vi]来实现的，每次"涨落"也必然带来系统的结构性演变。这个特点反映在空间上，体现为在发展过程中社区空间边界的变迁以及空间结构的变化。

对动态性的研究涉及到对事物发展的长期观察，由于时间条件并不具备，因此本研究只能够通过两方面工作来获取信息，一是现场采访租户和管理人员；二是阅读文献。通过汇总归纳获得的信息，对案例的边界和结构状况进行绘制，并试图区分不同的发展阶段。研究有如下三个发现（图29）：

首先，大部分案例的物理边界都比较很模糊。产业功能都是从局部入驻（如沙坡尾的滨水区、曾厝垵的村口和主街、大芬村的主街、798艺术区的几处厂房）再慢慢扩张。在发展的任一阶段，其物理边界都很难被明细地确定，只能够通过新入驻的功能在既有空间结构占据的位置来确定。当然，我们这里所说的"边界"，并不是这些案例作为"创意园区"的边界（这个边界往往是被后来的规划所确定的），而是指城区内具有产业特色的区域。

其次，边界呈动态性发展。随着创意社区的发展，边界往往是慢慢向外扩张的。在几个案例中，这个趋势都比较明显。各个发展阶段的边界都不尽相同。

最后，案例的各阶段发展都呈现一定结构性特征，但这些特征在发展过程中会发生改变。例如，大芬村的发展一开始是沿外围道路，但后期的发展则以主街为中心。798艺术区中三个主厂房各自形成独立中心，随后慢慢有合并趋势，同时贯穿中部的主要道路成为形成空间轴。沙坡尾从沿湾东侧的节点空间，慢慢发展成环湾多节点的空间结构，并且大学路成为平行滨水空间的另一条空间主轴。曾厝垵的空间结构则一直是沿着主要的街道发展，从点状发展成线状，到最后形成网络结构。

The regeneration of creative community is a dynamic process, in which its spatial structure constantly changes. This is an a prominent feature of the development of complex systems. According to the self-organization theory, the development of complex system is realized by "fluctuation" [vi], and "fluctuation" brings about the structural evolution of the system. This feature is to be reflected into space, in terms of its change of boundary and internal structure (Fig. 29).

Generally, the research on dynamic changes requires long-term observation. Nonetheless, only two resources are accessible for collecting data on the selected Chinese cases in this research due to the limiting condition. One is on-site interviews with original residents, tenants and administration staffs. The other is relevant documents. By summing up the information obtained, the boundary and structure of the case are outlined in our drawings, in which we seek to distinguish the distinct stages of development. Here come three findings of the research:

First, the physical boundaries of most cases are vague. Industrial functions first enter some selected areas and then gradually expand from them (such as the waterfront at Sha Po Wei, the village entrance and the main street of Zeng Cuo An Village, the main street of the Da Fen Village and several factories in the 798 Art Zone). At any stage of development, the physical boundaries are difficult to be defined, and can only be traced by the locations where buildings are occupied by the new functions. Of course, when we say "boundaries" here, we do not mean that the boundaries for "creative industrial parks" (which are often defined by subsequent planning), but rather refer to areas with industrial characteristics within urban areas.

Secondly, the change of boundary is a dynamic process, which evolves with time. As creative communities develop, the boundaries tend to expand outward slowly. In several cases, this trend is evident. Boundaries vary from one stage of development to another.

Finally, each stage of the development of the cases has some structural characteristics, but these characteristics will change during the development process. For example, the Da Fen Village developed initially along a peripheral road, but later it took the main street at the centre. The three major workshops in the 798 Art Zone form their own independent centers, and gradually merge together in its later stages of development. At the same time, the main roads through the middle of the district become the spatial axis. From the spatial nodes along the eastern side of the bay, the Sha Po Wei has gradually developed into a multi-node spatial structure around the bay, and the Daxue Road has become the secondary spatial principal axis of the parallel to waterfront space. The spatial structure of the Zeng Cuo An Village has been developed along the main streets, from a centralized form to a linear one, until network structure is finally formed.

沙坡尾 / Sha Po Wei

2001　2007　2018

曾厝垵 / Zeng Cuo An

1995　2005　2018

798 艺术区 / 798 Art Zone

2008　2012　2018

大芬村 / Da Fen Village

2008　2012　2018

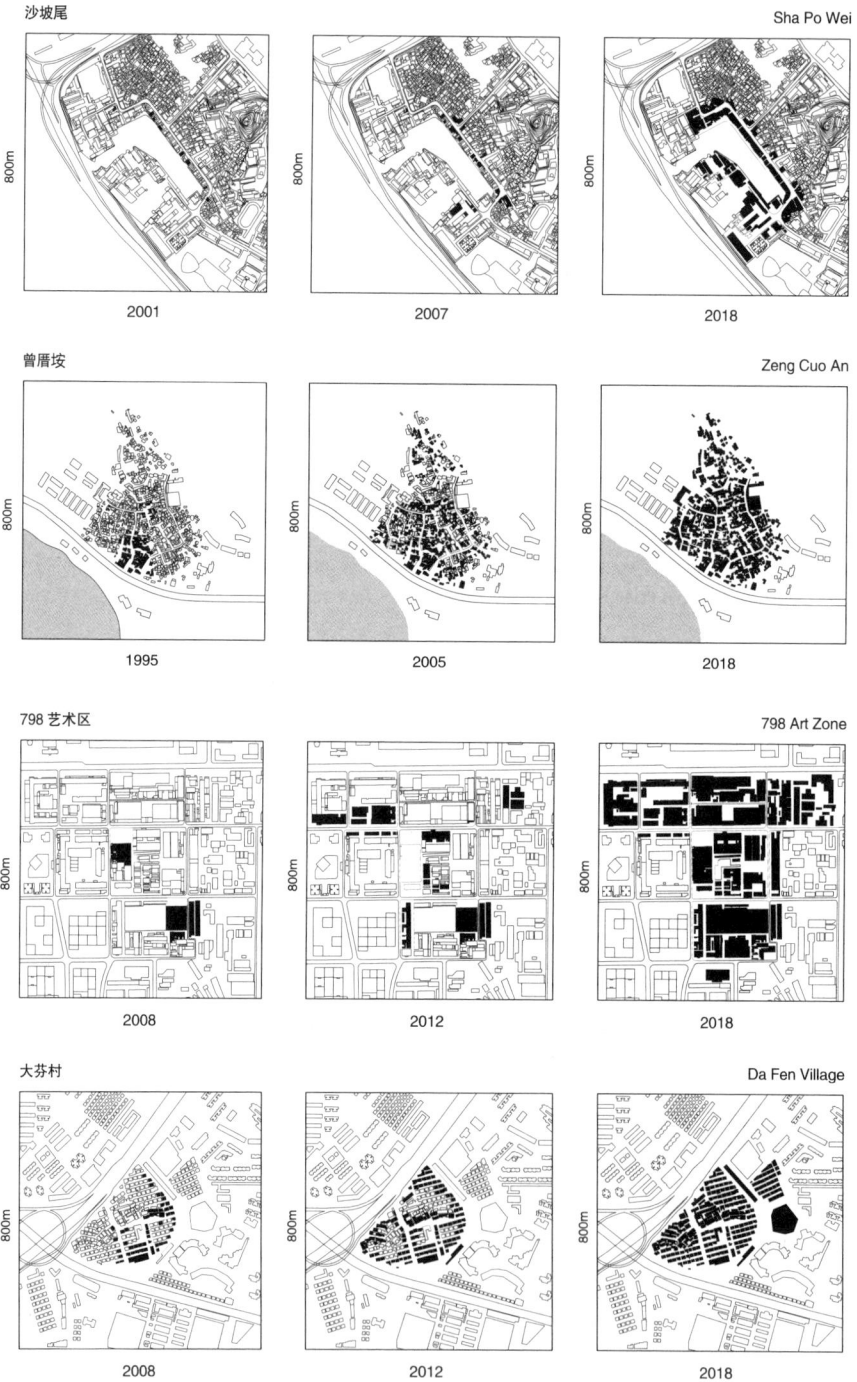

图 29 沙坡尾、曾厝垵、798 艺术区和大芬村的空间结构演进
Fig. 29 Spatial evolution of Sha Po Wei, Zeng Cuo An, 798 Art Zone and Da Fen Village

4.4　空间与产业的关联
ASSOCIATION BETWEEN SPACE AND INDUSTRIES

创意社区的空间形态需要反映社区各个利益相关者之间的关系，因此它必然具有很强的"关联性"特征。功能是反映使用状态的最重要指标，因此我们在这里将功能的空间布局进行绘制。我们还运用了一种分层的画法，这是基于这样的观察结果，即功能的分化在创意城区中是复杂的，垂直方向上分层是一个重要的特征。

我们将功能分为三类，第一类是直接面向消费的商业服务业，如各类纪念品和服饰商店以及餐厅、酒吧、咖啡厅等，这些功能在各个案例中占的比例都非常高，而且随着案例知名度的提高此类功能的增长非常迅速。第二类功能是创意产业，进一步划分为主产业和延伸产业。例如，在大芬村案例中的主产业是油画制作产业，包括整个产业链条中的油画原材料生产（画框、画布、画具等）、各类油画制作作坊、工人宿舍、仓储、画廊和博物馆。延伸产业是与油画产业链不直接关联的其他产业，如旅馆、媒体和设计公司。在798案例中主导产业是艺术产业，具体包括画廊、博物馆、艺术家工作室等。延伸产业则为广告、设计、服装等公司。曾厝垵案例中主导产业是特色民宿酒店。在沙坡尾案例中为文化展览相关的功能，但是相较于充分发展的服务业并不突出。第三类功能是城区既有功能。在其他功能的入驻后既有功能都还有一定的留存，与新进驻的功能产生关联。如在798案例中指的是原来的制造业企业、仓储和工人宿舍，在大芬村、曾厝垵和沙坡尾案例中则是原有城区的居住和相应的服务和附属设施。

从分析的结果上看，面向消费的服务业基本占据主要公共空间如街道的周边（如沙坡尾的内港岸线、曾厝垵的几条主街及主要节点、798艺术区的几处广场和大芬村的主街），这个规律符合服务业也以人流为导向的布局规律。创意产业同样比较依赖公共空间，因为好的服务设施以及公共空间的氛围是吸引创新人群的重要原因，但由于承担租金能力上低于商业服务业，它们也会在接近公共空间的第二层面分布，或者在第一层面的上部（如临街建筑的一层或二层为商业服务业，上部楼层为产业）。城区的其他功能则在空间的竞争中选择远离公共空间的位置（图30）。

总的来说，功能分布呈现如下规律。第一，很难在产业布局上看到明显的分区特征，而是呈现有机生长的状态，产业功能和非产业功能的关系（即所谓的"产城融合"）也是基于这个状态展开的。第二，公共空间影响功能和产业布局。三类功能都有临近公共空间布局的趋势，其中商业服务业和创意产业对公共空间的依赖最明显。第三，从房屋的功能混合度上看，临近公共空间房屋功能混合度往往较高，远离公共空间的房屋往往功能比较单一。

The spatial form of the creative community is expected to reflect the relationship among the various stakeholders, for which spatial layout of different functions is the most decisive indicator. Therefore, we work out the spatial layout of functions by employing a layering approach on the basis of the discovery that functional differentiation is complex in creative urban areas and that vertical layering is an outstanding feature.

We divide different functions into three categories. The first category is the consumption–oriented commercial functions, such as various souvenir and clothing stores, restaurants, bars, coffee shops, and so on. These functions account for a relatively high proportion in each case. And they are growing quickly as the popularity of the places in the cases increases. The second category is the creative industry, which is further broken down into the main industry and the extending industry. In the case of the Da Fen Village, for example, the main industry is the oil painting industry, which includes the production of oil painting raw materials (frames, canvas, painting tools, etc.), various oil painting workshops, workers' dormitories, warehouses, galleries and museums in the entire industrial chain. Extending industries are the others that are not directly related to the oil painting industry chain, such as hotels, media and design companies. In the 798 Art Zone, the main industry is the art industry, including art galleries, museums, artists' studios, and so forth. The extending industry is advertising, design, clothing and other companies. In the Zeng Cuo An Village, the main industry is boutique hotel. In Sha Po Wei, the main industry is that which serve for the cultural exhibition and the related activities. The third category is the remaining functions in the urban areas. In nearly all cases, after creative industries settle down, the existing functions are surviving to a certain extent. For example, in the 798 Art Zone, it refers to the original manufacturing enterprises, warehouses and workers' dormitories, while in the Da Fen Village, the Zeng Cuo An Village and Sha Po Mei, it refers to the remaining residence and ancillary facilities (daily shops, schools, markets etc.) in the urban area.

According to the analysis, the consumption-oriented service basically occupies major public spaces like the sides of the streets (such as the inner harbour shoreline at Sha Po Wei, several main streets and main nodes of the Zeng Cuo An Village, several squares of the 798 Art Zone and the main streets of the Da Fen Village). This law is in line with the flow of people, which benefits the business. Creative industries are also more dependent on public spaces, since high quality services and the atmosphere in public spaces are irreplaceable attractions to innovators. However, their ability to pay for space is lower than commercial services' in most situations, their location may also be close to the second level of the public space, or above the first level (for example, the first or the second floor of a street building is for business services, and the upper floors for creative industries). In contrast, the other functions which depend less on public space of the city are to choose the place away from this area in the competition of space (Fig. 30).

In general, the tendencies of functions layout are as follows. First, it is difficult to see obvious zoning characteristics in the industrial layout, but a characteristic of organic growth. The industrial and non-industrial function relationship is also found based on this characteristic. Second, the public space affects the function layout strongly. All the three categories of functions tend to be located near the public space, among which it is the commercial service and creative industry whose dependence upon the public space is most noticeable. Third, in terms of the function mixture, the buildings related to public space have a higher function mixture degree compared to those staying away from the public space.

三层及以上
Third Floor and Above

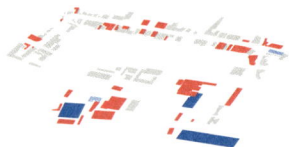

二层
Second Floor

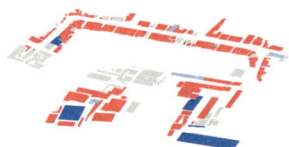

一层
First Floor

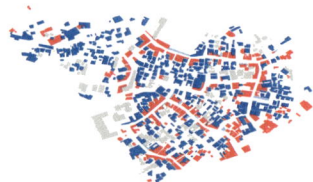

公共空间
Public Space

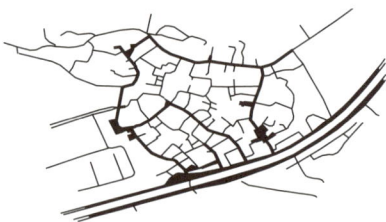

主要产业
Main industries

延伸产业
Extending industries

商业服务业
Commercial functions

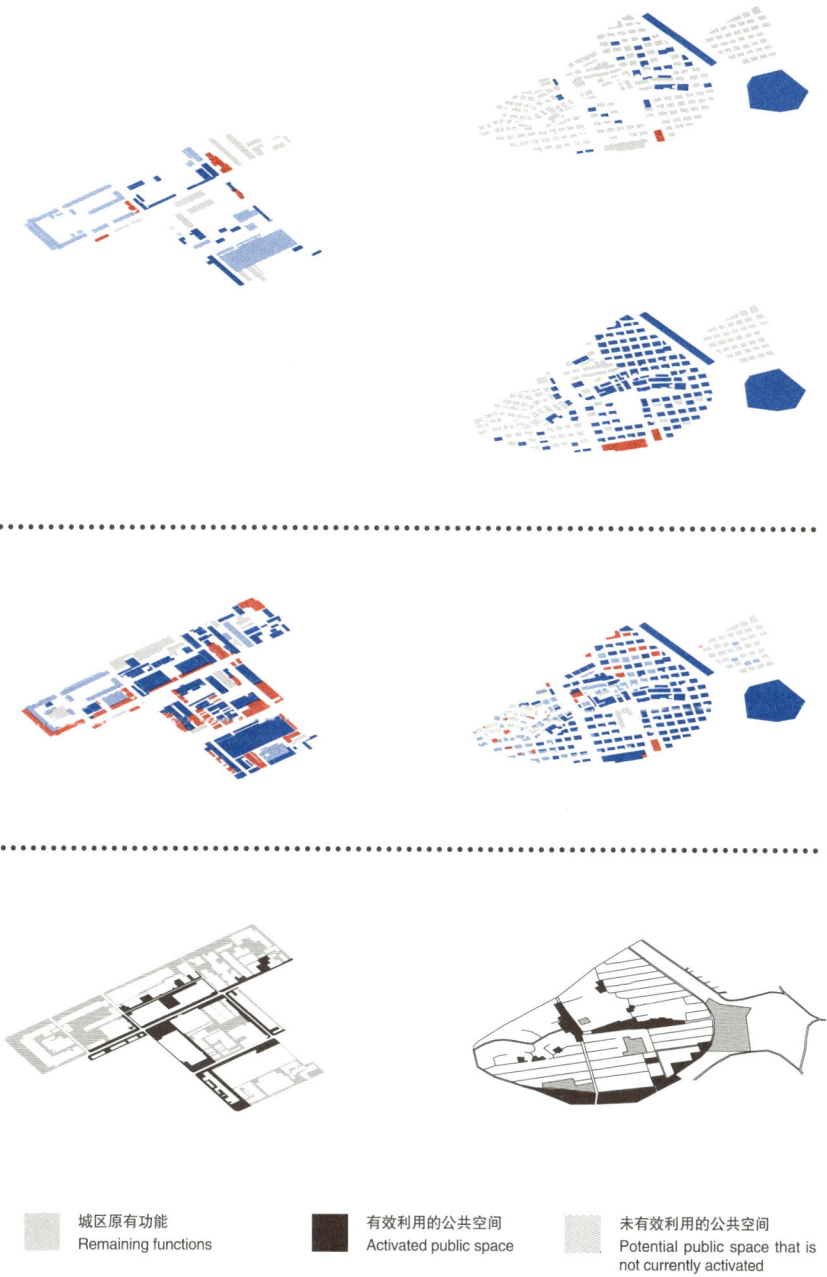

城区原有功能 Remaining functions	有效利用的公共空间 Activated public space	未有效利用的公共空间 Potential public space that is not currently activated

图 30 沙坡尾、曾厝垵、798 艺术区和大芬村的产业空间布局
Fig. 30 Distribution of industries in Sha Po Wei, Zeng Cuo An, 798 Art Zone and Da Fen Village

4.5 空间的类型化特征
TYPOLOGIES OF SPACE

每个创意社区，都有令人印象深刻的建筑类型，无论是798艺术区中的多跨单层厂房、两个城中村案例中的独立多层民宅建筑，还是沙坡尾中的沿街连排建筑或是田子坊中的里弄住宅。我们的研究发现，既有建筑的空间类型对功能再利用的可行性和可能性的影响是巨大的，这种影响最终决定了城区是否适合某种类型的产业入驻；而不同空间类型在同一案例中的分布（大多数情况下是既有的空间条件决定的），也从某种程度上决定了案例中产业分布规律（图31、32、33、34)。

就空间类型本身而言，比较有代表性的是大芬村和曾厝垵的小尺度独立式多层建筑，特别适合小型产业入驻，也同时适合其他的功能如住宅和商业，在很多地方呈现出多种多样的功能组合方式。沙坡尾的多层联排式建筑（指共用侧墙），也同样具有适应多种功能的优势，且具有公共性更强的底层（因为骑楼的存在），以及向两侧相邻空间单元进行功能扩张的可能性。在798艺术区中，多跨单层厂房空间则特别适合展览功能，同时可以兼容一些商业零售业，但对于住宅的兼容性就比较差。

此外，不同的建筑空间类型不仅决定了功能再利用状况，也决定了其使用室外空间的方式和对之进行管理的强度，进而影响了公共空间的布局和品质。某些建筑类型为了更好地利用公共空间，也对自身进行改造，如打开外墙和扩建建筑物的低层部分等。比如说曾厝垵和大芬村案例的独立式多层建筑对外部空间的利用就有较大的需求，从而促进了各案例中主要公共空间及其结构的形成。沙坡尾案例滨水一侧的建筑通过打开底层建筑的外墙，在滨水一侧形成商业街。798艺术区的多跨厂房也尝试打开部分外墙做商业运营，同时激活这些商业设施的前区空间。但是，也正因为这些外墙不具有被全部打开的可能性，导致这个案例的外部空间并未获得像其他案例般成功。

Every creative community has one or more impressive types of architecture, whether it is the multi–span single–storey factory building in the 798 Art Zone or the independent multi–storey villagers' housing in the case of two urban villages. Our research found that the space characteristic of existing buildings has a tremendous impact on the reuse feasibility by industries, which ultimately determines whether the urban area is suitable for certain types of industries. The distribution of different types of buildings in the same case, to some extent, determines the distribution of industries in the case (Fig. 31,32,33,34) .

As far as the typologies of buildings itself is concerned, the small scale independent multi–storey buildings in the Da Fen Village and the Zeng Cuo An Village are representative. They are particularly suitable for small enterprises, while also suitable for other functions such as residential and commercial. In many places, there appear a variety of ways to combine functions. Likewise, the multi–storey row building with a shared side wall at Sha Po Wei has the advantage of adapting to various functions, and has a better publicity in its ground floor (on account of the existence of the arcade), as well as the possibility of functional expansion to the adjacent space units on both sides. In the 798 Art Zone, multi–span single–storey factory space is particularly befit exhibition functions, and can be compatible with some commercial retails, but relatively poor for residential compatibility.

In addition, different typologies of buildings not only determine the reuse of their functions, but also determine the use of outdoor space and the intensity of the management, accordingly affecting the layout and the quality of public space. In order to make improved use of the public space, some types of buildings have also modified themselves, such as opening the outer walls and expanding the low–rise parts of the buildings. For example, the use of outer space in the independent multi–storey buildings of the 2 urban villages has a higher demand, thus promoting the formation of the major public space and its structure. The building on the waterfront side of the Sha Po Wei forms a commercial street on the waterfront side by opening the outer walls of the ground floor building. The multi–span workshops of the 798 Art Zone also attempted to open parts of the exterior walls for commercial operation, and at the same time to activate the front space of these shops. However, it is precisely because that these walls are not possible to be fully opened, the external space in this case was not as successful as in other cases.

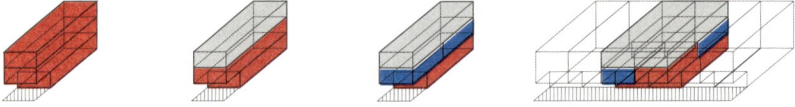

图 31 沙坡尾里沿街城市住宅的功能再利用和扩张模式
Fig. 31 Function combination and expansion model in street-side apartment housing in the case of Sha Po Wei

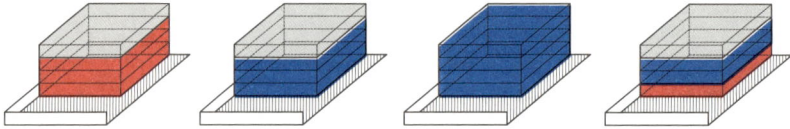

图 32 曾厝垵里村民住宅的功能再利用和扩张模式
Fig. 32 Function combination and expansion model in villager houses in the case of Zeng Cuo An

商业服务业
Commercial functions

创意产业
Creative industries

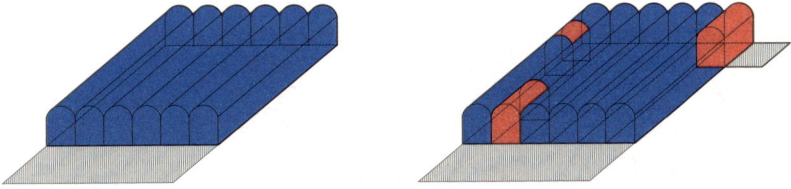

图 33 798 艺术区里连续多跨厂房的功能再利用和扩张模式
Fig. 33 Function combination and expansion model in multiple–span factory buildings in the case of 798 Art Zone

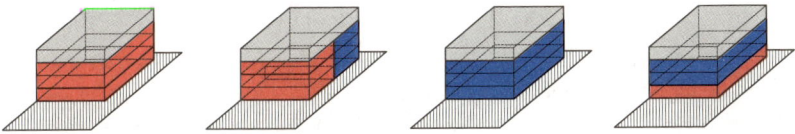

图 34 大芬村里村民住宅的功能再利用和扩张模式
Fig. 34 Typology of Buildings in Da Fen Village

居住
Residential function

被激活的公共空间公共空间
Activated public space

59

5 创意社区案例
CASES OF CREATIVE COMMUNITIES

5.1　案例分析的方法和内容
CONTENTS AND METHODOLOGY OF CASE ANALYSIS

这个章节展示的五个创意社区的案例，是国内近十多年来涌现的最有代表性的创意社区，它们是：北京 798 艺术区、深圳大芬村、厦门沙坡尾、曾厝垵和上海田子坊。选择这五个案例的原因在于如下一些方面：首先，它们都是在城市中心区或其附近的既有城区基础上，通过自下而上地改造既有城区，逐渐形成今天的格局；其次，它们占领的城市区域都比较大，且与周边的城区关系非常密切。城区内留存了较多既有功能和人群（如原住民），具备成为"社区"的基本条件。此外，案例都有比较鲜明的主导产业，主导产业在城区的产业生态中占据主要部分，并成为其他产业的基础。最后，这几个创意社区都经历了较长的发展历程，其空间形态已经呈比较稳定的状态，为本研究提供了较好的基础。

案例选择还注意突出了案例之间的差异性。从所在城区的特点上看，城中村（大芬村和曾厝垵）、历史街区（田子坊和沙坡尾）和工业厂房（798 艺术区）都具有彼此不可替代的特点，揭示出创意产业在和城区物质空间结合时呈现出的多样性。从创意产业类型上看，坐落于城市中心区的田子坊、沙坡尾和曾厝垵所呈现的产业特色（以面向旅游的创意产业为主）就和坐落于中心区边缘的 798 艺术区和大芬村的产业特色非常不同，他们面临的产业发展机会和危机也各不相同。此外，很多案例所在城区都带有一定历史文化特色，这些特色能够帮助创意社区实现一定的知名度。而文化遗产在城区的自发更新中，到底是得到了加强还是面临威胁，也是一个需要考察的问题。

正如我们已经在"选址规律"(4.1) 中提出的, 除了经济原因之外, 空间特质是创意社区在这些城区发生的重要因素。这就是本章通过各种空间分析要着重揭示的内容。分析在下面这些方面展开：

1. 区位和资源
2. 城市肌理
3. 周边城市公共空间网络
4. 产业类型及其空间分布分析
5. 典型建筑类型及其更新方式

我们兼顾分析的客观性和表达的艺术性。对于城区空间系统、房屋信息和空间使用信息, 我们通过现场采集（一些重要数据通过逐户调查）的方式来收集。所有的分析图, 都采用统一格式和方法绘制, 在数据可信的基础上将五个案例放在相同的平台上进行呈现和对比。我们也坚信, 艺术具有它穿透事实, 直面本质的力量。因此, 建筑布局、公共空间以及典型建筑类型这几个方面, 我们采用平面绘本的方式, 呈现人和空间的密切关系, 尤其是隐藏在建筑立面背后那些微妙的空间使用。

The 5 examples of creative communities presented in this chapter are the most representative creative communities that have emerged in China in the past decade. They are, the 798 Art Zone in Beijing, Da Fen Village in Shenzhen, Sha Po Wei and Zeng Cuo An in Xiamen, as well as Tian Zi Fang in Shanghai. The reasons for electing these cases are as follows. First of all, they take shape in or near a city's central area by transforming existing urban areas through bottom-up regeneration. Secondly, they occupy large urban areas and incorporate in close relationship with the surrounding urban areas. In these urban areas, remaining functions and staying people are joined with new ones, which give the urban areas the identity of "communities". Thirdly, most creative communities have a dominant industry, which takes the leading role in the industrial eco-system of the areas and acts as a basis to supporting other industries. Finally, these creative communities have experienced a long process of development, as a consequence, their spatial form has been relatively stable, which confirms a reliable and valid base for this study.

The differences between the cases are also highlighted here. Viewing from characteristics of the urban areas, the urban village (Da Fen Village and Zeng Cuo An Village), the historical blocks (Tian Zi Fang and Sha Po Mei) and former manufacturing district (798 Art Zone) all have their irreplaceable attributes. This reveals the diversity of creative industries when they are associated with the physical space of urban areas. In terms of the types of creative industries, Tian Zi fang, Sha Po Wei and Zeng Cuo An Village, which are located in the city's central area, display industrial characteristics (primarily tourism-oriented creative industries) that apparently differ from those of 798 Art Zone and Da Fen Villages, which are situated in the peripheral area of the city. Also, they run into different opportunities and crises in the process of industrial development. In addition, many cases have a certain historical and cultural traits, which, of course, help the creative community achieve a certain degree of popularity. However, whether cultural heritage has been strengthened or threatened in the bottom-up regeneration process is yet a concern that needs to be examined.

As we have pointed out in the "spatial factors of site selection"(4.1), apart from economic factors, spatial factors are important reasons for creative communities to occur in these urban areas. This is the essential content of this chapter through a variety of spatial analysis. The analysis is developed in the following areas:

1. Location and resources
2. Urban fabric
3. Public space net in its surroundings
4. Categories of industry and their spatial layout
5. Building typologies and their regeneration methods

We take the objectivity of the analysis and the artistry of expression into unbiased consideration. Regarding the information about urban spatial systems, building states and their spatial use, we collect them through the field investigation (important data even via house–to–house survey). All the analysis diagrams are drawn in a uniform format with uniform mothodology, in order to present and compare the five cases on an equivalent platform on the basis of high data–credibility. Firmly, we also believe in art's power which can penetrate into facts from outside to reach the very center of the essence. Therefore, in the aspects of the architectural layout, public space and typical architectural patterns, we use the way of graphic depiction to show the connection between people and space, especially the subtle application of hidden space behind a building's facade.

沙坡尾

5.2 | SHA PO WEI

沙坡尾位于厦门老市区的南部滨海区域，是早期厦门港的一部分。这是一处弧形的海湾，沙滩连成一片，历史上有"玉沙坡"美称。因为渔港的存在，沿岸很多渔民住宅和渔业房屋聚集，形成一定规模，并与老城连成一片。2005 年以后，渔业逐渐衰落，房屋出现一定闲置的情况。一些特色的艺术产业开始入驻，逐渐形成规模，给予区域极大的特色，吸引很多游客和市民前来观光和体验。2015 年，政府针对沙坡尾的景观整治规划出台，规划措施包括对港区的清淤、岸线拓宽以及对西岸的一些空间整理。同时进行的对现存渔业和渔民的迁出运动，招致极大争议。总的来说，沙坡尾仍是厦门老市区最有特色的城区之一，其产业、空间和建筑的多样性令人叹为观止。

Sha Po Wei is located in the southern coastal area adjacent to the old urban area of the City of Xiamen and is a part of the early Xiamen Port. Surrounding the arc–shaped bay, the beach is interconnected, the laudatory name "jade sand slope" is thus given. Because of the existence of fishing ports, many houses related to fisherman and fishing industry gather along the coast, and are connected with the old city. After 2000, the fishing industry has declined gradually, and the houses appeared to be idle to a certain extent. Some art and culture industries began to settle since then, giving the region great characteristics, attracting tourists and citizens to come for sightseeing. In 2015, the municipal government launched a set of planning, whose measures include the dredging of the port area and the widening of the shoreline, as well as some spatial rearrangement of the southern shore. The move–out of the fishing industry and fishermen caused some controversy. In general, Sha Po Wei is still one of the most characteristic urban areas in the old urban area of Xiamen, and its diversity in industry, space and architecture is astonishing.

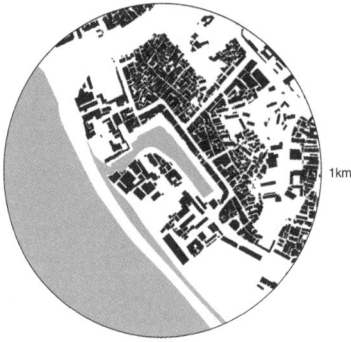

城市空间肌理图

Figure ground plan map

卫星图

Satellite map

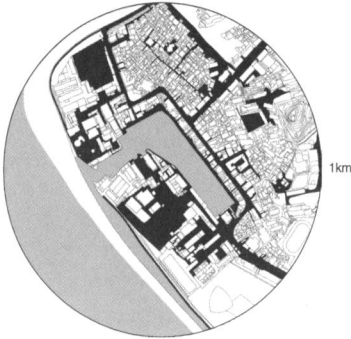

路网与广场

Roads and plazas

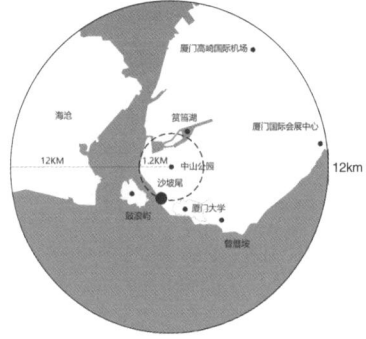

区位图

Location

　　沙坡尾地处厦门老城区边缘地带，和老城区的空间系统相连，结构性特征类似，即以地块外部街道为空间主轴，地块内部的街巷阡陌交错。滨港区的步行带目前已经基本连成一片，串联各个节点，是区内的主要空间结构特征。大学路，是一条两侧有骑楼的特色街道，是平行于上述空间的第二条空间轴线，它一端与主城区相连，另一端连接厦门大学与环岛路，这条路是游客和外来人员进入城区的主要通道。

Sha Po Wei is located at the edge of the old urban area of Xiamen. It is connected with the spatial system of the old urban area and has similar structural characteristics, that is, the streets outside the large plot are the area's main spatial axis, while in the plot there are very dense net composed of sub-streets and lands between houses. At present, the main spatial structure is characterized by the long shoreline surrounding the harbor with many scattering urban nodes. Daxue Road, a characteristic street with historic arcades on both sides, is the second spatial axis parallel to the above one. It connects with the old city at one end and Xiamen University at the other. This road is the main passageway for tourists and outsiders to enter Sha Po Wei.

从以渔业为主导，变成以文化艺术为主导的产业构成，反映了城市产业变迁的总体趋势。艺术产业作为主导产业，受到旅游商业的挤压（这与不同产业租金承受能力差异的关系很大），体现为主导产业在总量中的占比不大。从空间布局角度，商业和主导产业占据滨水和沿街的主要空间，对公共空间的支持是很大的，也是目前沙坡尾公共空间活力旺盛的主要原因。

The shift from a fisheries-dominated industrial structure to an art & culture-dominated industrial structure reflects the overall trend of urban industrial change. As main industries, the culture & art industry is squeezed by commerce oriented to tourism (difference of rent affordability of different industries play a role here), which is reflected in the shrinking percentage of the leading industry in the total amount. From the perspective of spatial layout, commercial and main industries occupy the main space along the waterfront and along the street. They largely support public space, which leads to the exuberant vitality of the public space in Sha Po Wei.

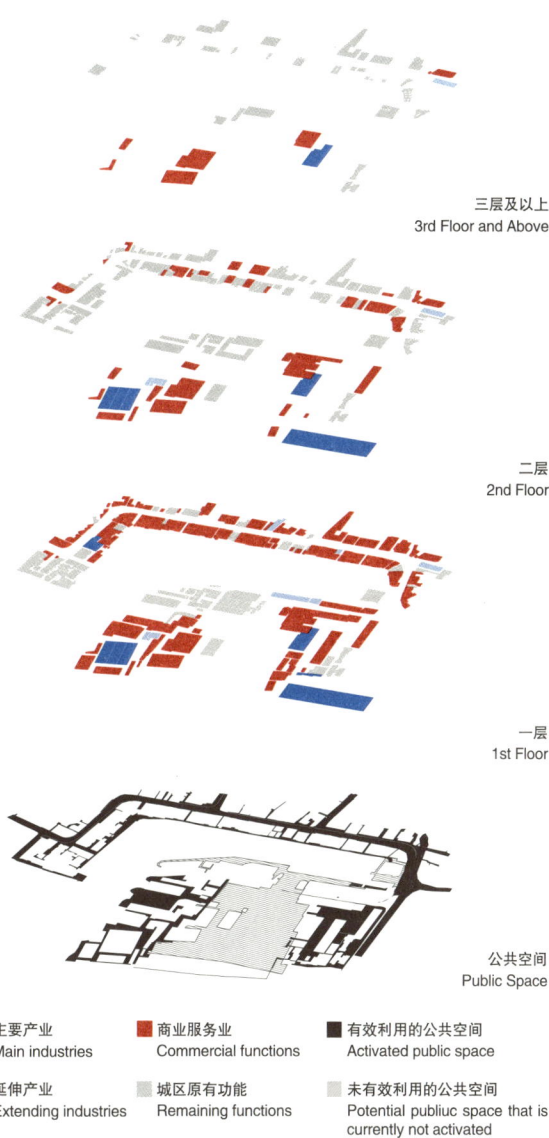

三层及以上
3rd Floor and Above

二层
2nd Floor

一层
1st Floor

公共空间
Public Space

■ 主要产业
Main industries

■ 商业服务业
Commercial functions

■ 有效利用的公共空间
Activated public space

延伸产业
Extending industries

城区原有功能
Remaining functions

未有效利用的公共空间
Potential publiuc space that is currently not activated

功能分类 Category of functions	具体功能及企业 Function description and enterprise	比例 Proportion
保留的城区原功能 Remaining functions	住宅及附属设施 Housing and affiliated facilities	44%
商业服务业 Commercial functions	餐饮店（111 家）Gastronomy (111) 零售店（65 家）Retails (65)	42%
主要产业 Main industries	文化艺术产业（7家） Culture & Art Industry (7)	12%
延伸产业 Extending industry	旅游服务（7家） Tourist service (7)	2%

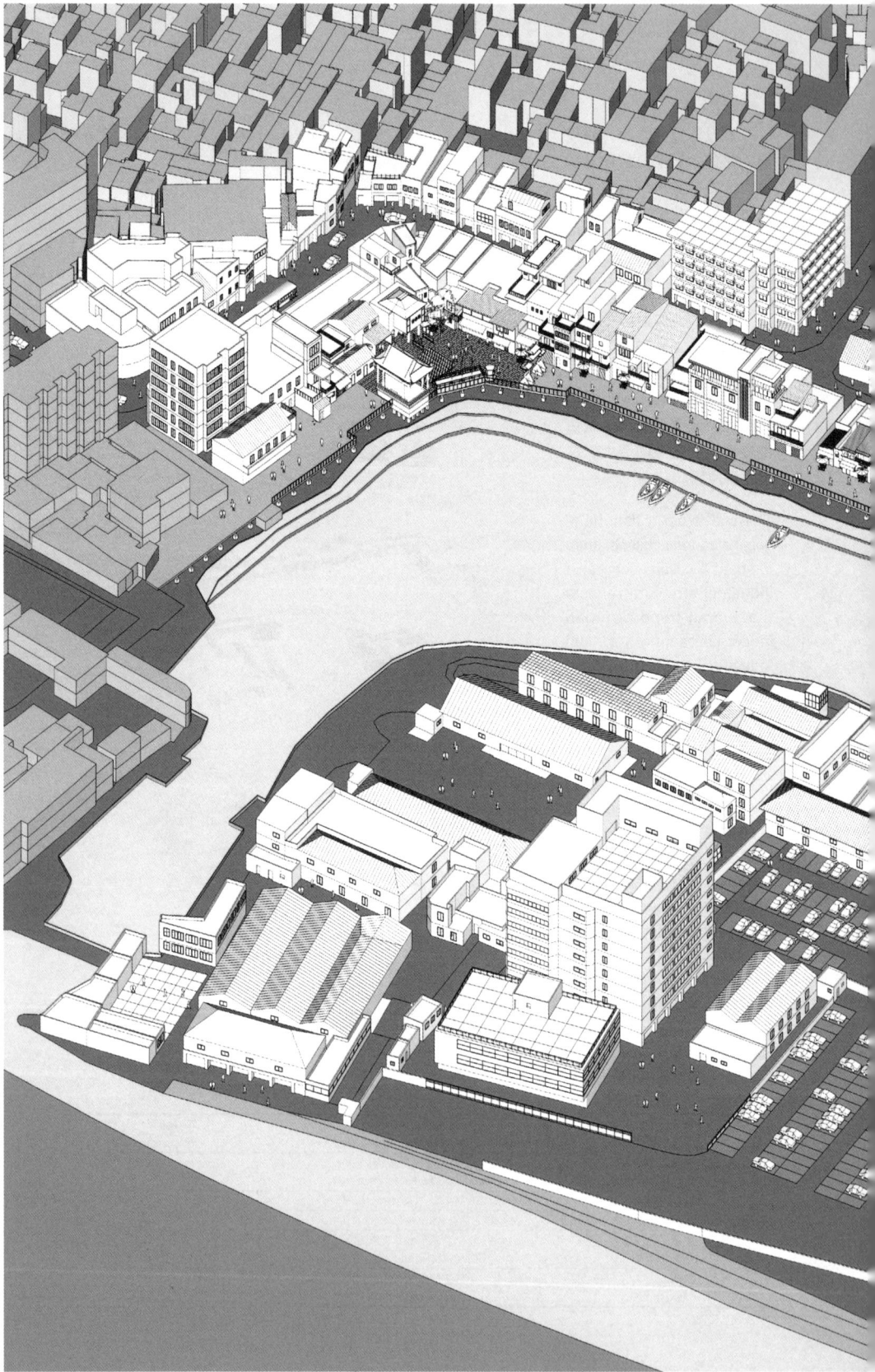

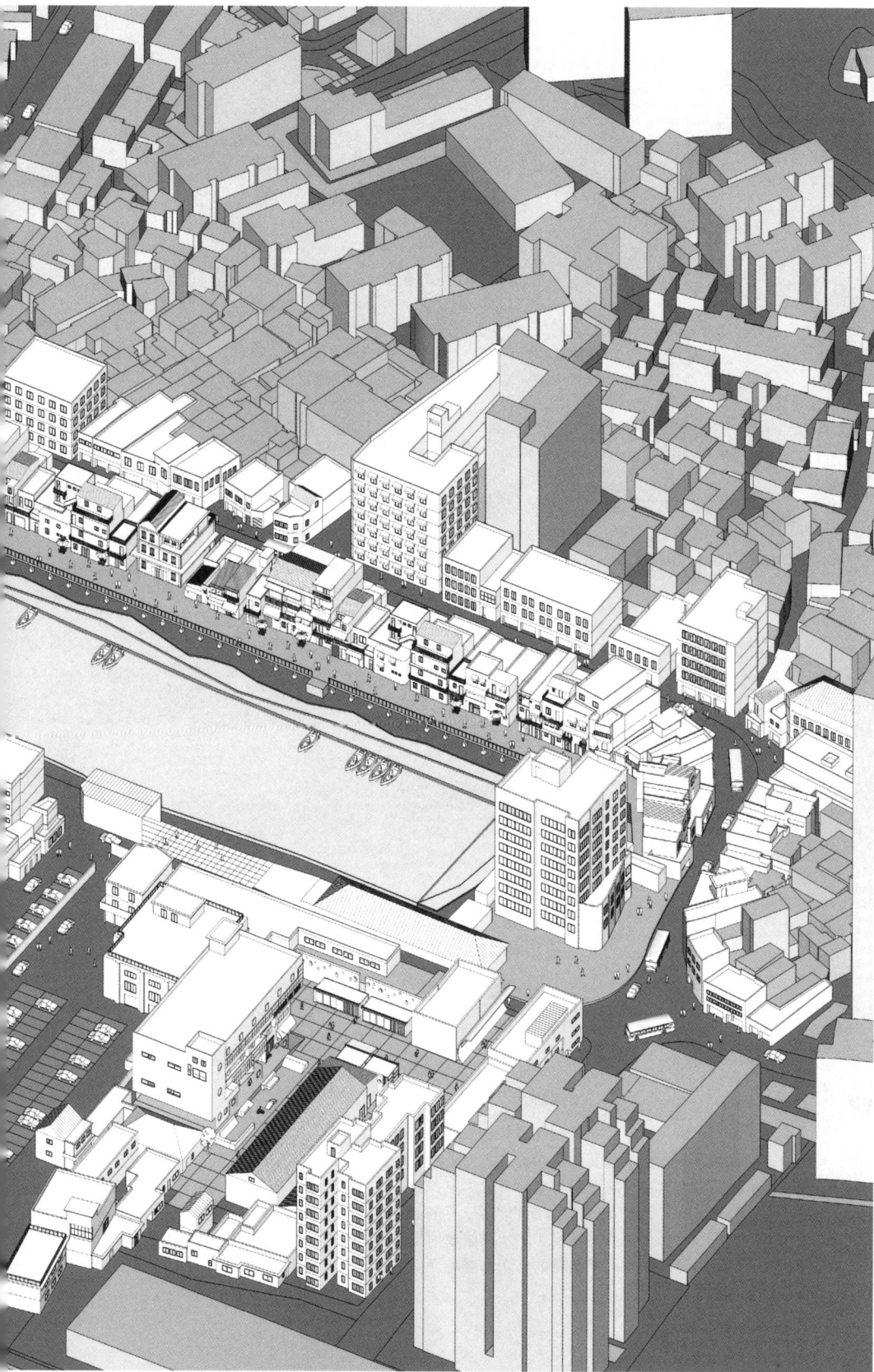

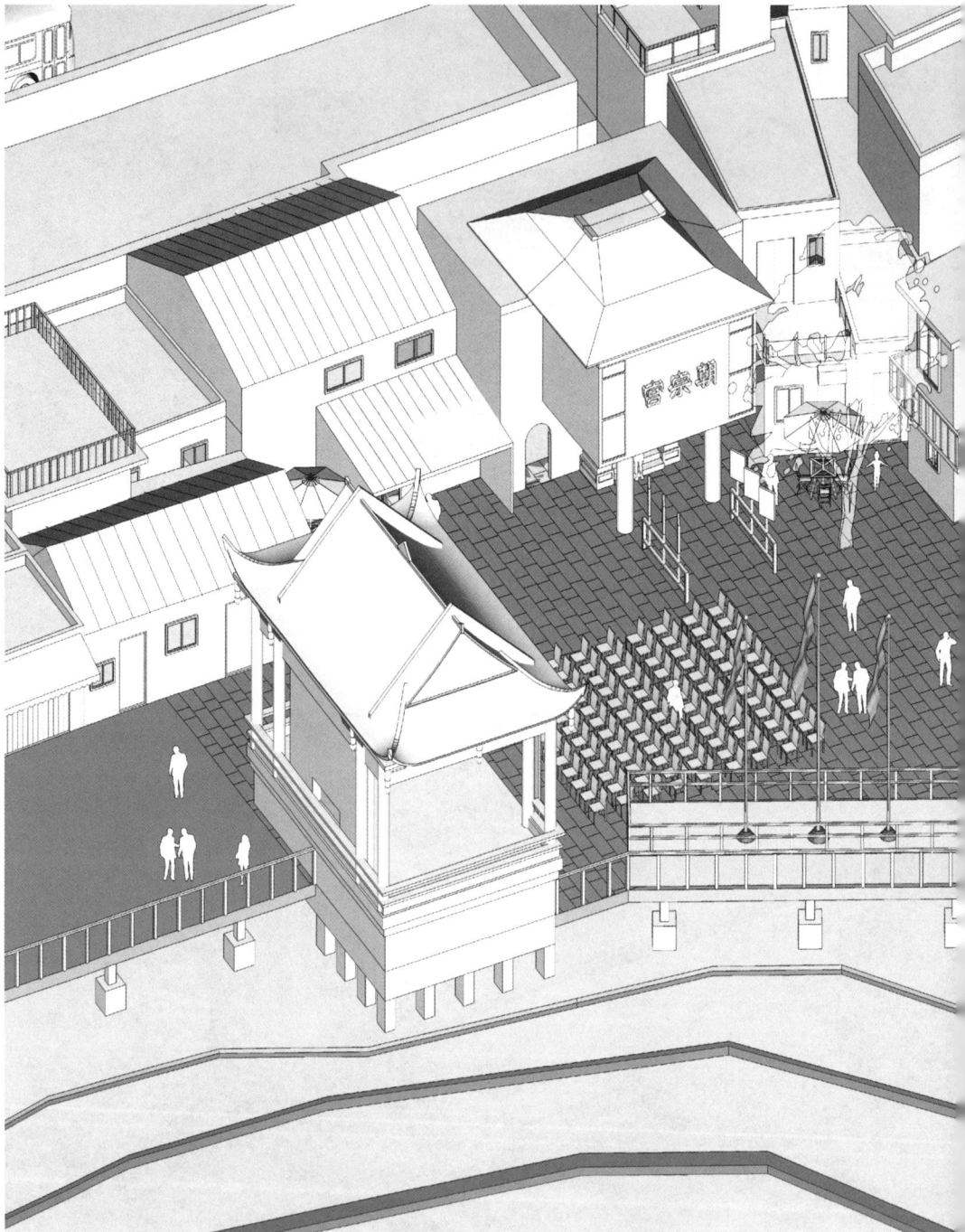

　　"避风坞"是城区最重要的空间要素，它是沙坡尾的一张名片。当人们从大学路的骑楼，穿过房屋间窄窄的一段小径，立刻被突然呈现在眼前的宏大景观所震撼。人们不仅赞叹渔港的风情，更是赞叹那么多商店、酒吧、工作室簇拥在一起形成的人文氛围。滨水岸线成为最受游客欢迎的场所，在这里他们不仅可以光顾风情独特的小店和咖啡厅，还能体验原汁原味的庙会和地方戏。

The inner harbor is the most important spatial element in this urban area, and it is a name card for Sha Po Wei. As people passed through the narrow paths from the Daxue Road arcade, they are immediately struck by the vast landscape that suddenly appeared in front of them. People praise not only the customs of the fishing port, and also such a humanistic atmosphere formed by the crowd of so many shops, bars, workshops. Of course, the waterfront is the most popular place for tourists. Here they can visit many unique shops and cafes, as well as become audience for temple fairs and local operas.

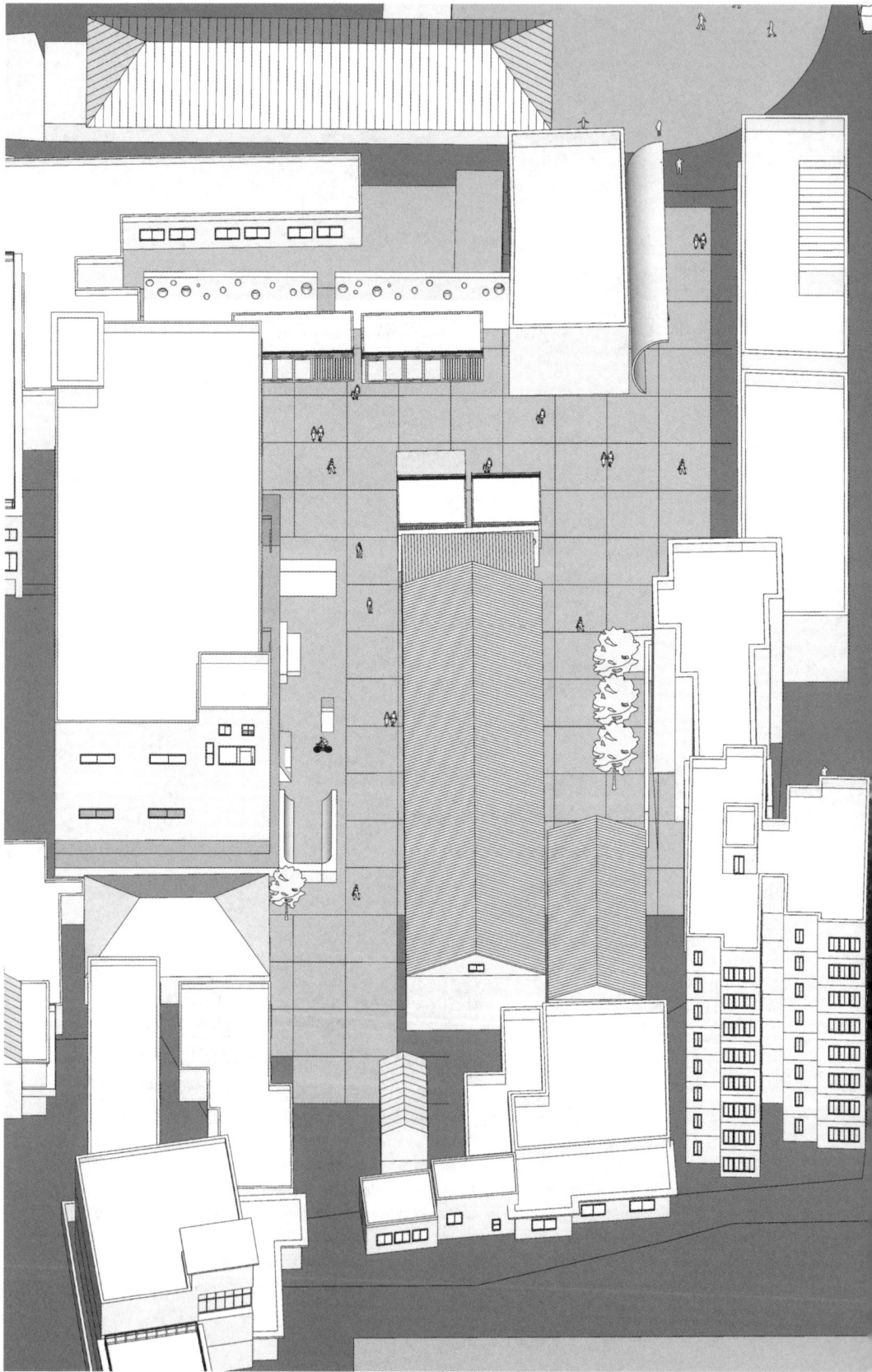

类型一

多层联排城市住宅

这些富有历史特色的城市住宅是沙坡尾最典型的房屋类型。面对大学路的一侧是骑楼，另一侧则临近避风坞。它的底层和二层常常被改造成餐厅、咖啡和特色商店，上面的楼层则作为工作室或者维持原有的住宅功能。当然，一墙之隔的另一个单元可能也加入空间改造，与它连成一体。这些房屋常常会增加楼层以扩大租户的利益，尽管此类改建行为往往是被禁止的。这些空间单元的两侧都是非常活跃的公共空间。

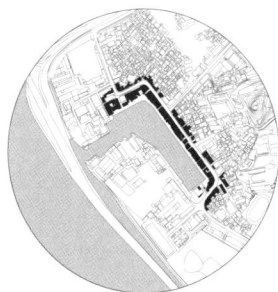

Type 1

The multi-storey row-housing

These are buildings rich in historic characteristics. They face to the Daxue Road in one side, where arcade is a strong identity, and the inner harbor in the other side. The lower floors are often converted into restaurants, coffee shops and specialty stores, with the upper floors serving as studios or maintaining their original residential functions. Of course, adjacent neighbored unit may also join in the space transformation to form larger groups. These houses often have additional floors to expand tenants' interests, although such alterations are often prohibited. These units are flanked by very active public spaces on both sides.

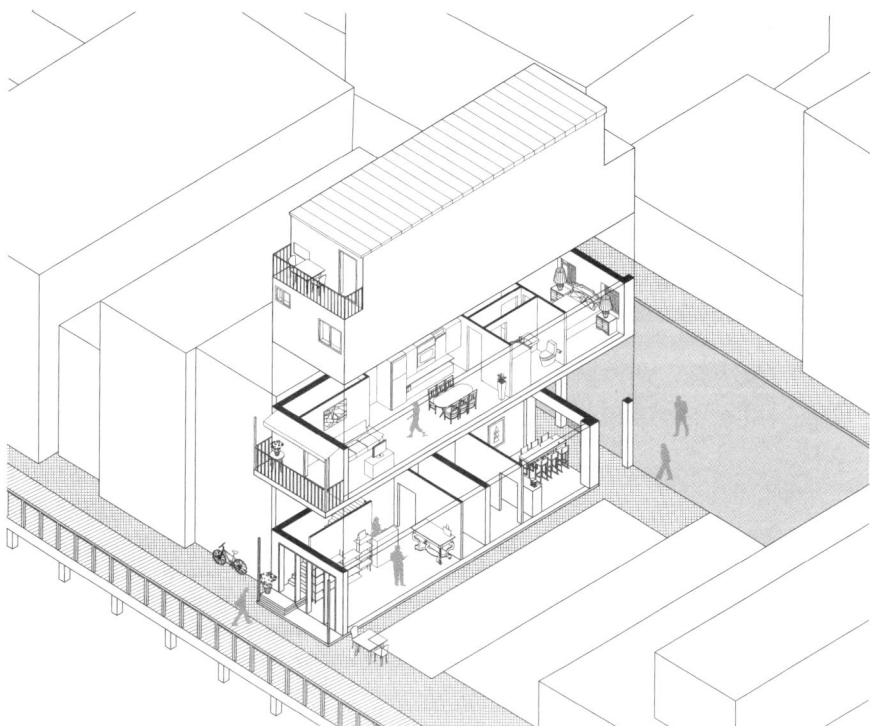

类型二

多层通用厂房

沙坡尾南岸有一些多层通用厂房，它们是原来的产业留下的闲置房屋。现在被灵活地划分成多样的空间，满足创意办公、展览和艺术商店的使用。它们的立面会被简单地改造，但原始房屋的基本特色仍然会保留。这些厂房的前区往往是艺术氛围浓烈的广场和街道。

Type 2

Multi-storey factory buildings

These are located mostly in southern bank of inner harbor. Being left idle by the manufacturing industries, they are now flexibly divided into various spaces for creative offices, exhibitions and art shops. The facade will be simply modified, but the basic features of the original house are preserved. The front areas of these plants are often squares and streets with strong art atmosphere.

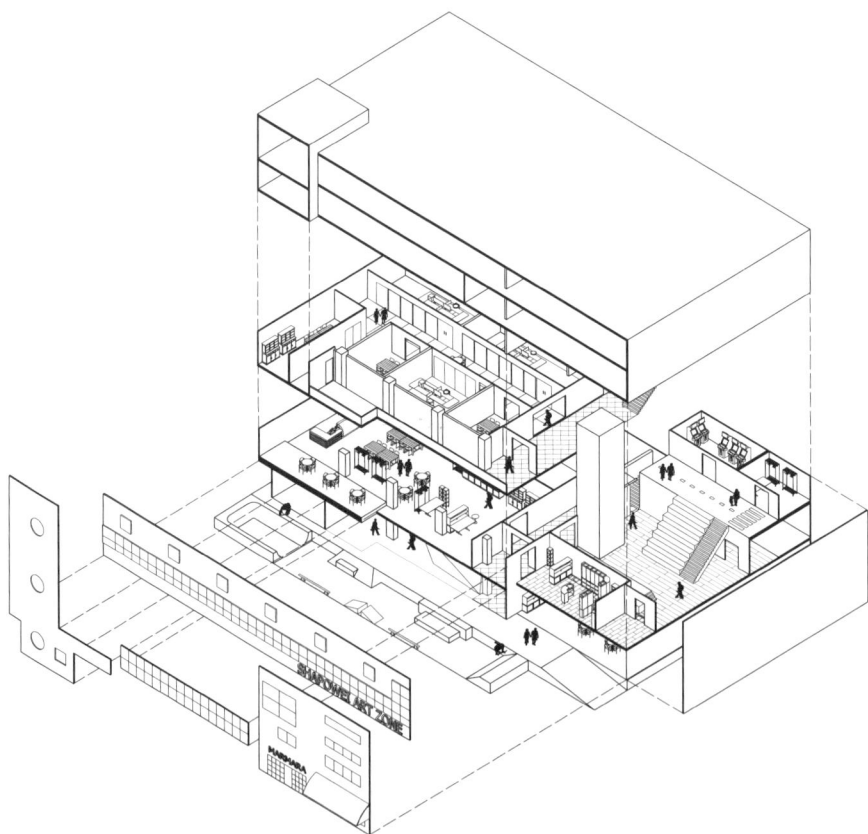

类型三

独栋单层仓库

钢结构屋顶仓库建筑是很有特色的。很高的室内空间让它们特别适合举办艺术展览。当然，餐厅和共享办公也很喜欢这样的空间，它们往往会被进一步改造，将大空间分割成适宜使用的小一点的区域，或者在大空间内增加楼层以扩大使用面积。

Type 3

Single–storey warehouse

These warehouse buildings are very characteristic by their steel–structured roof. The high interior space makes them particularly suitable for art exhibitions. Of course, restaurants and shared offices also like them, which is often further modified by dividing the large space into smaller areas suitable for use, or adding floors within the large space to increase the size of the use area.

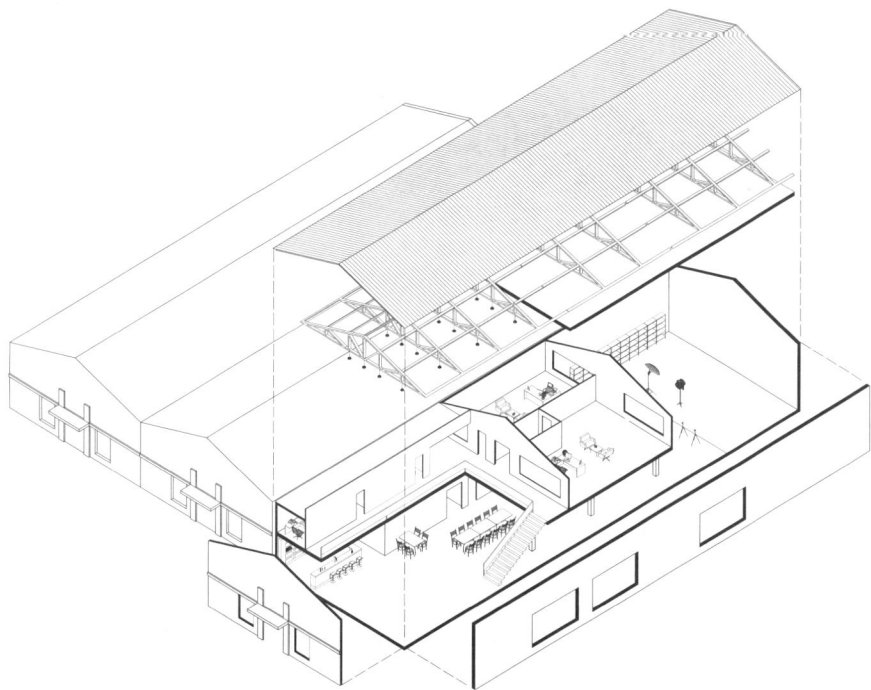

曾厝垵

5.3 | ZENG CUO AN

曾厝垵坐落于厦门本岛的南部，紧邻环岛路和沙滩。它是历史悠久的渔村，也曾经是厦门华侨聚集的地方。今天，它被称为"中国最美的渔村"，是著名的旅游景点。游客蜂拥而至来体验渔村的风情，他们当中的很多人，还会住几晚。精品酒店产业迅速地发展起来，成为这里的重要产业，渔民也纷纷从经营渔业转向经营酒店。宅基地和房屋就是村民创业的资本。村里很多小街小巷，被改造成步行商业街。最有意思的是，很多村民把房屋的一部分专做经营后，他们自己仍然愿意住在这里，他们的孩子还在这里的学校上学。庙会、祭祖和一些其他民间活动仍然定期进行。在强大的商业化影响面前，社区仍然在延续。

Zeng Cuo An is located in the southeast of Xiamen Island, close to Round–Island Road and the long beach. It is a fishing village with a long history, and it was once a place where overseas Chinese gathered in Xiamen. Today, being a great touristic attraction, it is praised as "the most beautiful fishing village in China". Tourists run to the fishing village, many of whom stayed for a few nights. Boutique hotel industry has developed rapidly, becoming an important industry here. Fishermen have also shifted from operating fisheries to operating hotels. Homestead and houses are the capital for villagers to start their own businesses. Many small streets and alleyways in the village have been transformed into pedestrian shopping streets. The most interesting thing is that many villagers still want to live here after they have managed part of their houses exclusively as hotel, and their children are still studying in the schools here. Temple fairs, ancestor worship and all other folk activities continue on a regular basis. In the face of commercial influences, communities continue.

城市空间肌理图

Figure ground plan map

卫星图

Satellite map

路网与广场

Roads and plazas

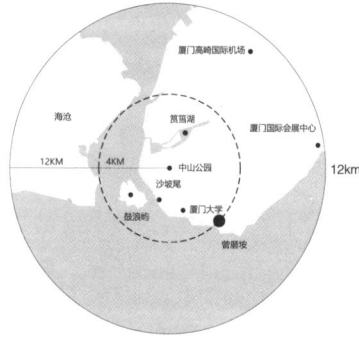

区位图

Location

曾厝垵的空间肌理在它的周边环境里是独特的——它几乎是以自我为中心的。除了向南侧打开，从环岛路和沙滩吸纳人流，它在其他的方向上都是封闭的。街巷的网络是它最重要的空间结构，面向环岛路的村口和贯穿南北的主街是最重要的场所，活力旺盛，各种商店和精品酒店也最多。当然，也有一些区域的游客稍微少一些（特别是那些远离街巷的地方），那里村民的居住环境就会更好一些。新入驻的产业和既有的社区的生活，是那么奇妙地结合在一起。

The urban fabric of Zeng Cuo An is unique in its surroundings, as it is almost self-centered by being closed in all directions except to the south to draw people from the Round-Island Road and the beach. The network of alleys is its most important spatial structure, in which the village entrance facing the Round-Island Road and the long main street running through the north and south are the most important and vigorous places, where various shop and boutique hotels are gathering. Of course, there are also areas where fewer visitors access (especially those far from the streets), where the villagers live in a better environment. The new industries and the lives of the existing residents are so wonderfully intertwined.

精品酒店是曾厝垵最重要的产业，它是村民收入的主要来源。除此之外，服务于旅游的商业也得到了很大的发展。有一个事实不该忘记，那就是艺术产业曾经是最早入驻曾厝垵的产业类型，而在高租金的压力下，这些产业现在已经难觅踪迹。留存的既有功能有很多，学校、宗庙、教堂等设施都得到保存，很多居民也还住在这里。

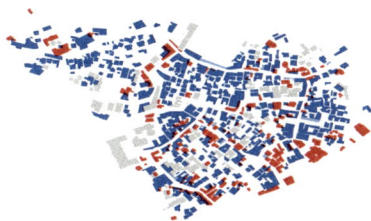

二层及以上
2nd floor and above

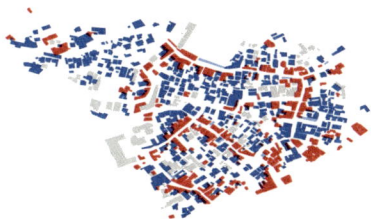

一层
1st floor

Boutique hotel is the most important industry in Zeng Cuo An, and it is the main source of income for the villagers. In addition, the service business related to tourism has also been greatly developed. One fact should not be forgotten, that is, the art industry was once the first type of industry to have settled in Zeng Cuo An. Under the pressure of high rents, these industries are now hard to find. There are many remaining community functions such as housings, schools, ancestral temples, churches.

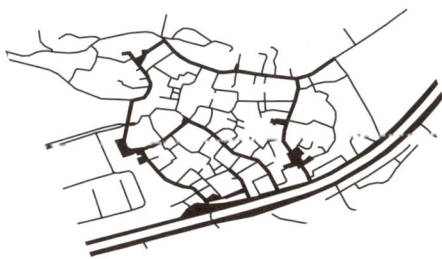

公共空间
Public space

■ 主要产业
Main industries

■ 商业服务业
Commercial functions

■ 有效利用的公共空间
Activated public space

■ 延伸产业
Extending industries

城区原有功能
Remaining functions

未有效利用的公共空间
Potential publiuc space that is currently not activated

功能分类 Category of functions	具体功能及企业 Function description and enterprise	比例 proportion
保留的城区原功能 Remaining functions	住宅及附属设施 Housing and affiliated facilities	14%
商业服务业 Commercial functions	餐饮店（194家）Gastronomy (194)	24%
	沿街摊位（215家）Street stalls（215）	
	纪念品商店（58家）Souvenir shop（58）	
	其他零售店（186家）Other retails (186)	
主要产业 Main industries	民宿酒店（190家）Homestay（190）	57%
延伸产业 Extending industries	办公（10家）Office（10）	11%
	艺术商店（10家）Art shop（10）	
	工作坊（8家）Workshops（8）	

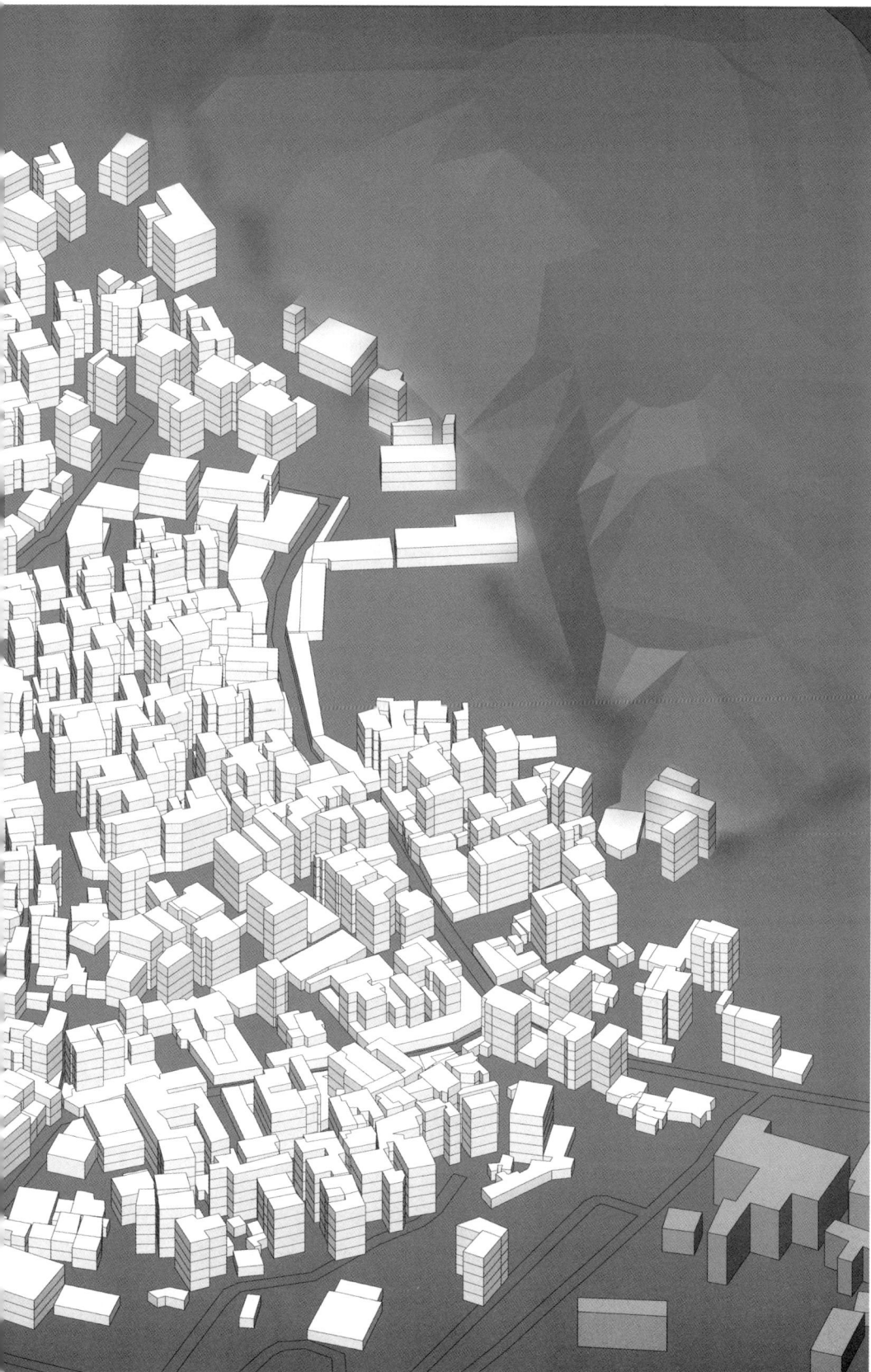

　　看似杂乱的村落空间，隐藏着巨大的丰富和多样性，这也许就是曾厝垵如此受游客欢迎的原因之一。宗祠、教堂、祖屋和新建设的多层村民住宅（大部分变成了民宿酒店）聚集在一起，形成一种奇特的和谐。在这里唯一可辨的空间要素，是横贯村子的主街。与环岛路相连的入口广场是经过一定规划的，似乎还有着某种规划赋予的"秩序"，但随着人们沿着主街往里走，往往就逐渐失去了方向感和明确的辨识性，不知不觉地参与到多样性的狂欢中去了。

The seemingly chaotic village space conceals great richness and diversity, which may be one of the reasons why Zeng Cuo An is so popular with tourists. Ancestral shrines, churches, ancestral houses and newly constructed multi-storey villager houses, most of which have become boutique hotels, come together in a curious harmony. The only discernible element of space here is the main street traversing the village from south to north. The entrance plaza connecting to the Round-Island Road is planned and seems to have some kind of "order" given by the plannin, but people often lose their sense of direction and clear discernment as they walk along further with the main street, unwittingly involved in carnival of variety.

类型一

多层村民住宅

这是曾厝垵最为典型的空间单元，大部分房屋都属于这种类型。该类型的房屋对酒店的功能具有极大的兼容性，也许是酒店产业在曾厝垵蓬勃发展的原因。基本的改造方式是，村民把围绕宅基地的院墙打开，把房屋以外的自用地开放，还对之进行简单环境改善以招徕游客的停留。房屋底层改造为商店和酒店大堂，上面的楼层改造成客房，留一部分面积供房屋主人居住。顶层和底层常常都会有加建，以扩大经营和居住面积，尽管该行为是被官方禁止的。

Type 1

Multi-storey villager houses

This is the most typical building unit in Zeng Cuo An. Most of the houses here belong to this type. The great compatibility of this type with the functions of hotel may have been the reason why the industry flourished in Zeng Cuo An. The basic method of transformation is that the villagers open the courtyard fences around the site and open the land outside the house for public. Simple environmental improvements are often made to attract tourists to stay. The ground floor of the house is converted into a shop and a hotel lobby, and the upper floors are converted into a guest rooms, leaving a portion of the floor area for the owner to live in. Both the top and bottom floors are often expanded to increase business and living space, although the practice is officially prohibited.

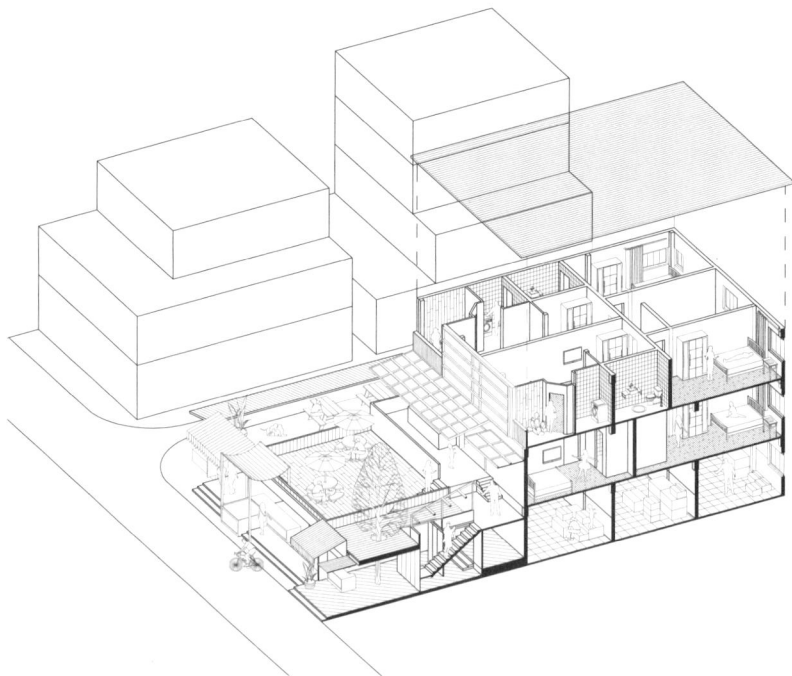

类型二

多层板式集合住宅

大部分属于这个类型的建筑也被改造成酒店。当然，因为不具有多层独立式村民住宅的特点，这里的酒店多是便捷式的、规模较大的连锁酒店，服务的是另一类的人群。有意思的是，由于房屋的规模略大，还出现多个酒店品牌共同经营房屋不同部分的情况。当然，建筑的底层多是被改造成商业。

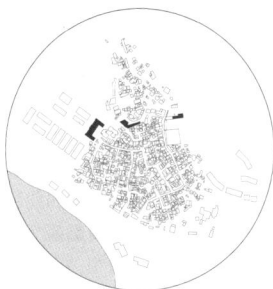

Type 2

Multi–storey collective housing in slab form

Buildings in this type have also been mostly converted into hotels. Of course, because it does not have the characteristics of multi–storey villager house, most of the hotels here are larger–scaled convenient hotels, serving another kind of visitors. Interestingly, due to the slightly larger size of buildings, it could happen that there are a number of hotel brands jointly operating same building in its different parts. Of course, most of the ground floor area of the buildings has been converted into a commercial uses.

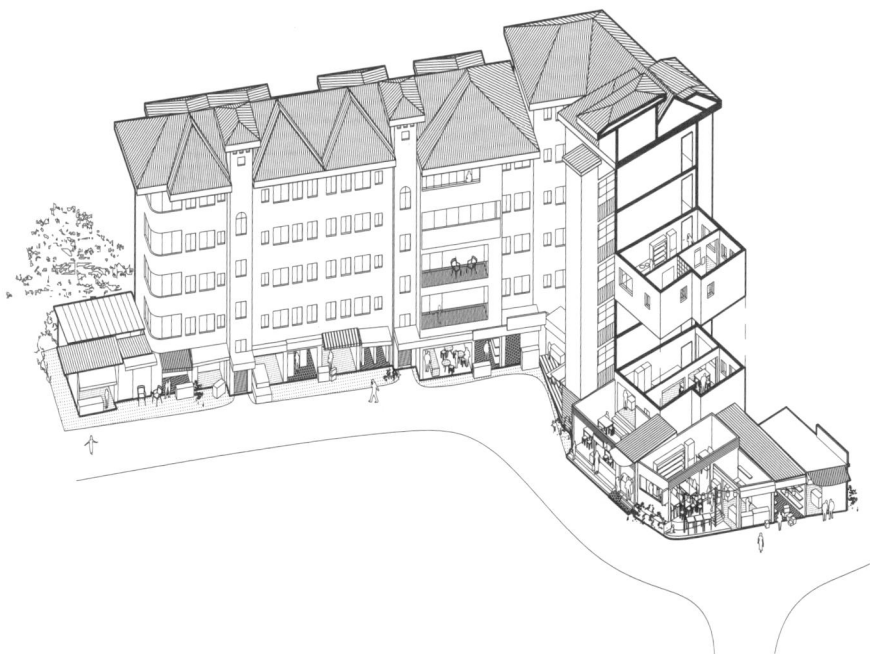

798 艺术区

5.4 | 798 ART ZONE

798 艺术区可能是中国最具国际影响力的艺术区。它那么频繁地出现于国际媒体，已经毫无疑问地成为中国当代艺术的最重要名片。令人惊讶的是，当代艺术和建于 50 年前的现代主义风格的厂房，居然如此和谐地共存，仿佛那些高敞厂房空间原本就是为当代艺术展览而打造一样。当我们回想起 798 艺术区在获得社会普遍关注之前曾经经历过的艰难的成长期，想起那些不同利益相关者之间的纷争，就会意识到城市发展里的自组织是如此的强烈。利益相关者的微妙作用至今还在对 798 艺术区的空间发展起着调节作用。如何让 798 进一步发展，保持它的产业多样性，并进一步提升城区的整体品质，是一个值得我们思考的问题。

798 Art Zone is probably the most internationally influential art district in China. It appears in the international media so frequently that it has undoubtedly become the most important business card of Chinese contemporary art. It is surprising that contemporary art and modernist buildings built 50 years ago coexist so harmoniously as if the open spaces had been created for contemporary art exhibitions with intention. When we recall to the difficult growing period of the 798 Art Zone before it gained the general attention of the society, and the disputes among the different stakeholders, we realize that the self–organizing characteristics of the urban development are so strong. The subtle role of stakeholders is still playing a regulatory role in the spatial development of 798 Art District today. Can 798 Art Zone be further developed with higher industrial diversity and enhanced overall quality of environment, is a question worthy of our consideration.

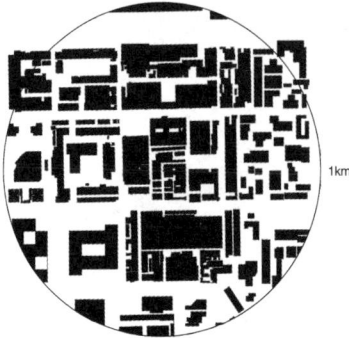

城市空间肌理图
Figure ground plan map

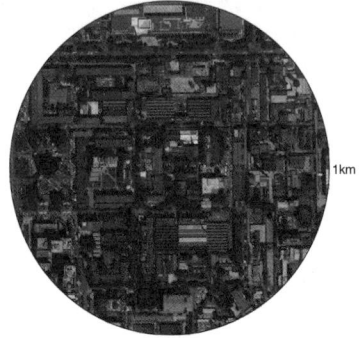

卫星图
Satellite map

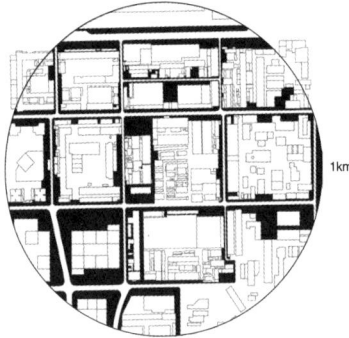

路网与广场
Roads and plazas

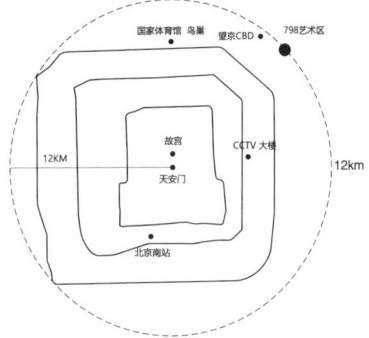

区位图
Location

在这连成一片的厂区，厂房之间的通道本是为了工业生产的流线而设计的。现在，一些通道变成了熙熙攘攘的道路，停满了汽车，和城市道路并没有太大不同。和几个最重要厂房连接的空间，被改造成了广场，经常举办一些艺术活动，他们成为区域的中心和主要节点。让我们设想一下，如果几条主要道路更适合步行、环境更舒适（而不是像现在这样成为大停车场），也许它们能够将整个区域的公共空间连成一片。

In this large contiguous factory area, the passageways between the factory buildings were originally designed for the streamline of industrial production. Now, some of the passageways have become bustling roads, full of cars, which are not much different from city's traffic roads. The spaces which are connected to several of the most important workshops have been transformed into squares, in which artistic events are held. They have become the centers and main nodes of the urban area. Let's imagine that, if a few main roads were more walkable and with a better environment quality (rather than a large parking place as they are now), they might be able to connect those public nodes into an entire network of public space.

798 艺术区的厂房，特别适合于举办艺术展览，因此艺术是最早进入这里的产业类型，像一些画廊和艺术家工作室。接下来，和艺术相关的其他产业，如媒体和设计产业，也纷纷进入，逐渐形成一个创意产业的聚落。此外，迄今为止，还有一些仓储、小型制造业和职工宿舍在遗留在区内。

Workshop buildings here are especially suitable for holding art exhibitions, so art is one of the first types of industry to enter here, such as some galleries and artists' studios. Later, other art-related industries, such as media and design, also entered, gradually forming a community of creative industries. In addition, to date, a number of warehouses, small-scale manufacturing and staff accommodation still remain in the urban area.

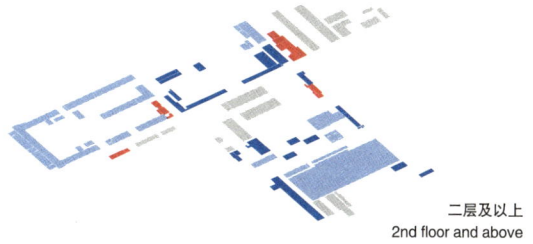

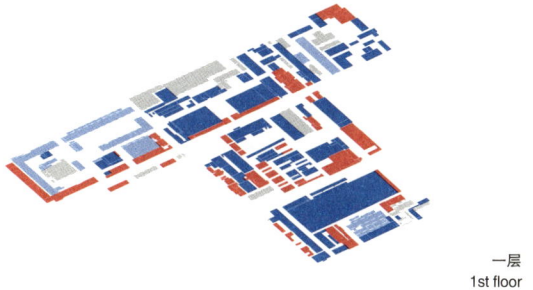

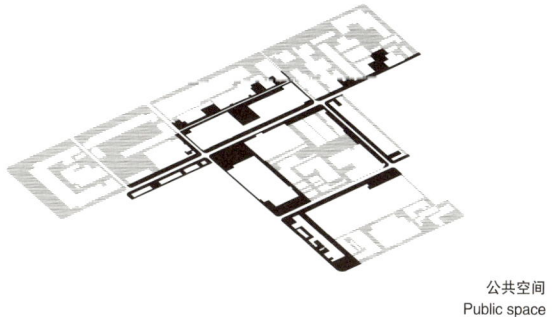

二层及以上
2nd floor and above

一层
1st floor

公共空间
Public space

■ 主要产业
Main industries

■ 商业服务业
Commercial functions

■ 有效利用的公共空间
Activated public space

■ 延伸产业
Extending industries

城区原有功能
Remaining functions

未有效利用的公共空间
Potential publiuc space that is
currently not activated

功能分类 Category of functions	具体功能及企业 Function description and enterprise	比例 proportion
保留的城区原功能 Remaining functions	工厂及附属设施 Factory and affiliated facilities	4%
商业服务业 Commercial functions	餐饮店（63 家）Gastronomy (63)	25%
	服饰店（36 家）Clothing shop（36）	
主要产业 Main industries	画廊（115 家）Gallery（115）	60%
	艺术中心（33 家）Art center（33）	
延伸产业 Extending industries	设计公司（27 家）Design company（27 家）	11%
	教育机构（16 家）Educational institutions（16）	

99

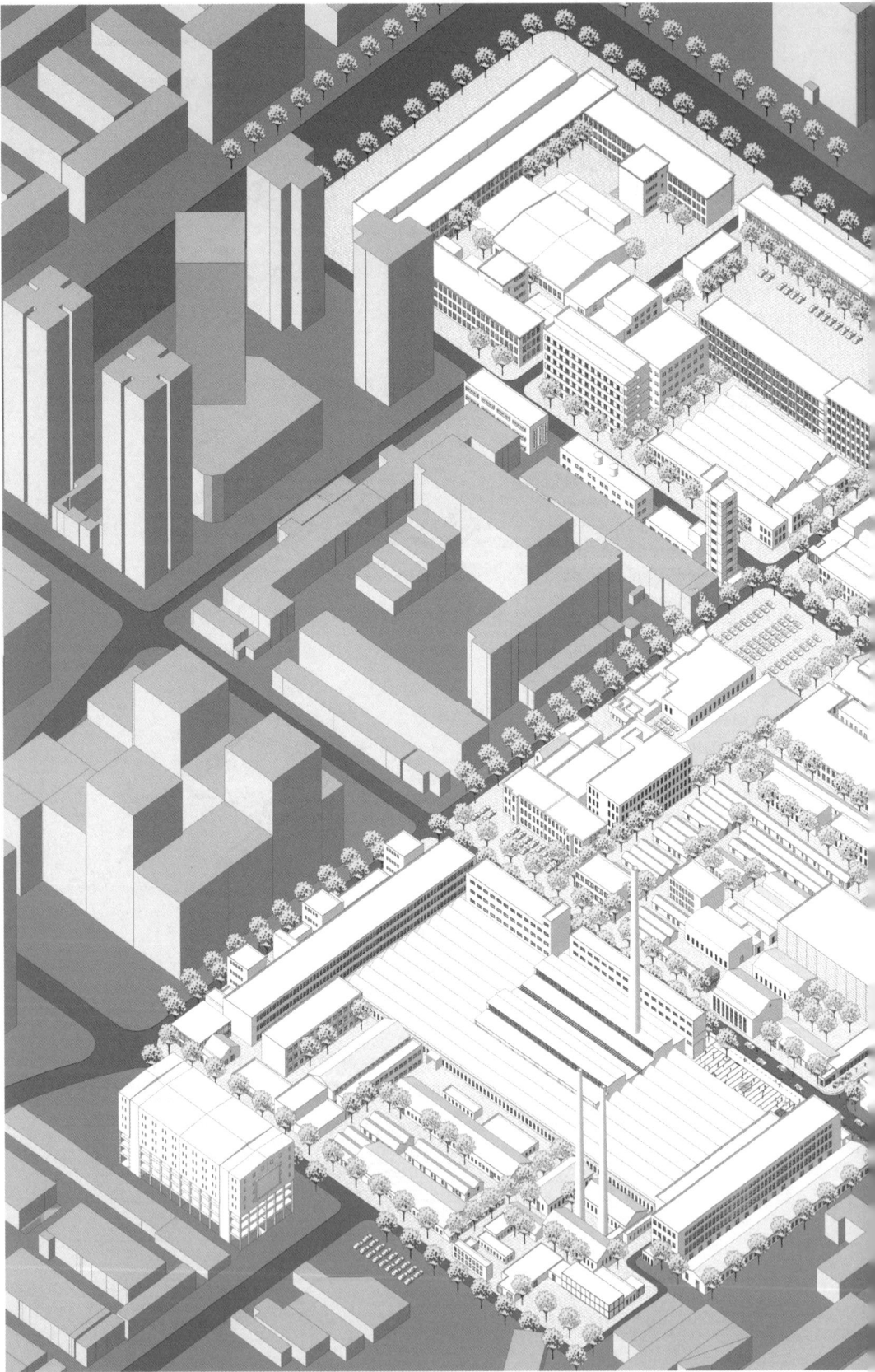

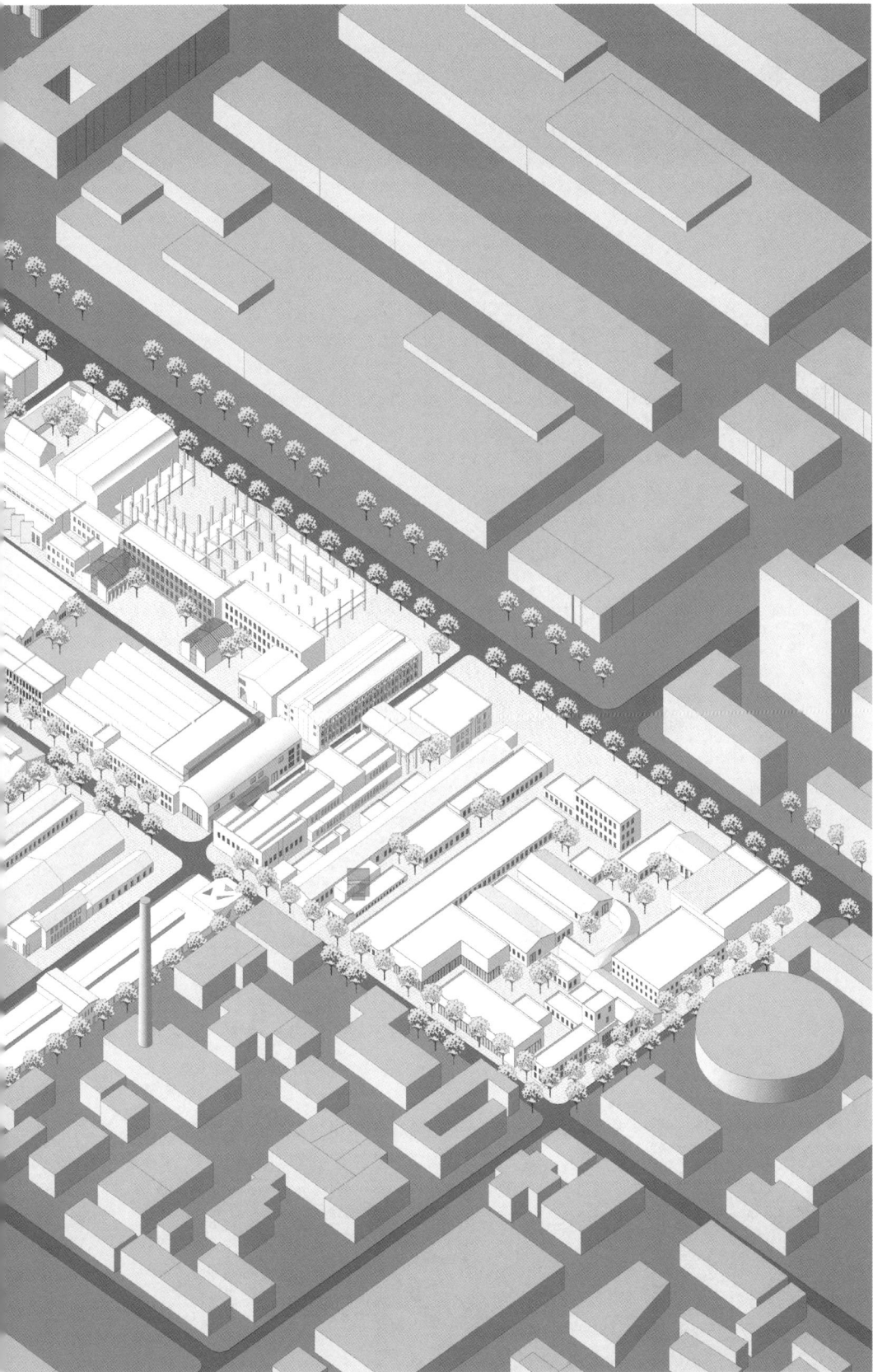

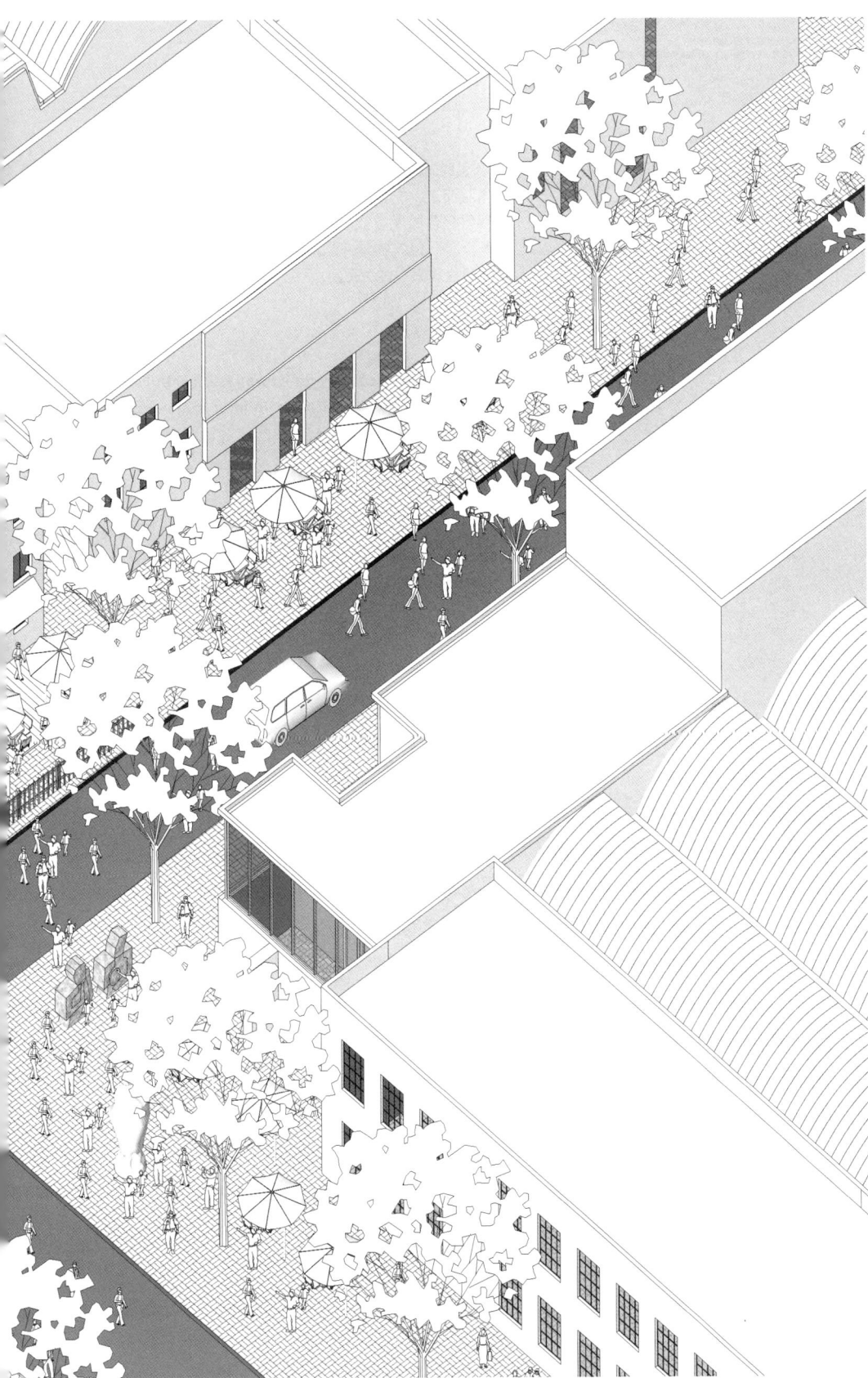

类型一

有条形天窗的多跨度单层厂房

这是最适合艺术展览的空间类型。柔和的天光从长窗倾泻而下，在宽大、高耸和明亮的空间中艺术展览可以获得极佳的观赏效果。除了艺术展览，也有少部分被用于办公。建筑物本身也具有很高的艺术价值，这也是很多观众前来观赏的目的之一。

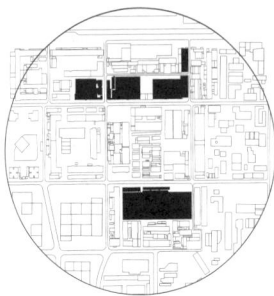

Type 1

Multi-span single-storey factory buildings with long roof windows

This is the type of space that is best suited for art exhibitions. Art exhibitions can be enjoyed in a borderless, tall and bright space, where soft skylight pours down from the long roof window, In addition to art exhibitions, a small number of them are also used for creative office. The building itself has a high artistic value, which is one of the attractions for many visitors.

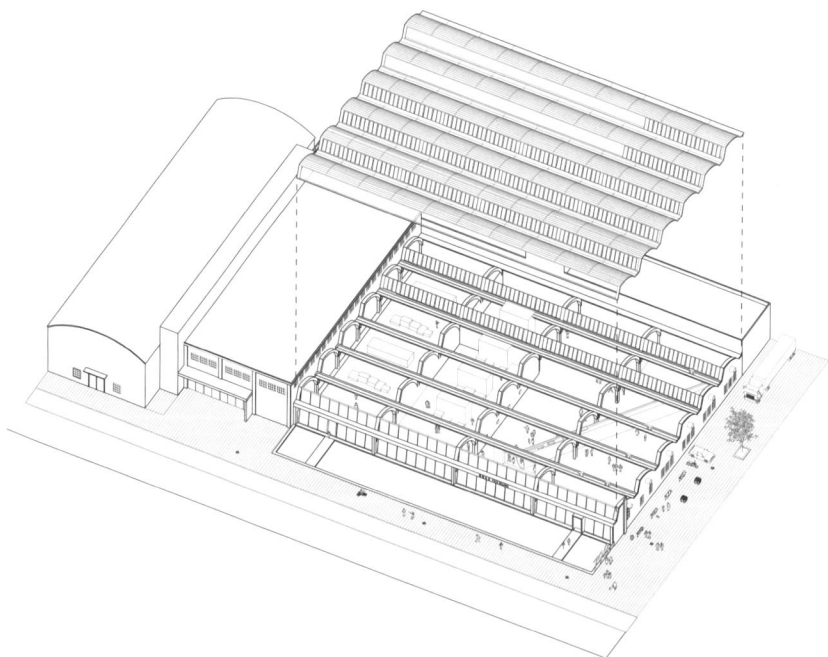

类型二

大型钢结构屋架单跨单层厂房

该空间类型也非常适合展览，因为空间非常夸大和高耸。室内可以进行夹层，以获得更丰富的空间效果和更多实用空间。一部分沿外部街道和广场的空间会被改造成服务游客的商业设施和餐饮店。

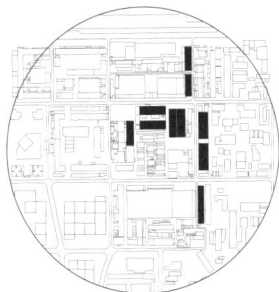

Type 2

Large single-span single-storey factory buildings with steel roof structure

They are also very suitable for art exhibitions, because the space is large and high. More floors or other building elements can be added to their interior space, for a better visual effect and advantage in use. Part of the space along the street or plaza are transformed into commercials and gastronomies serving tourists.

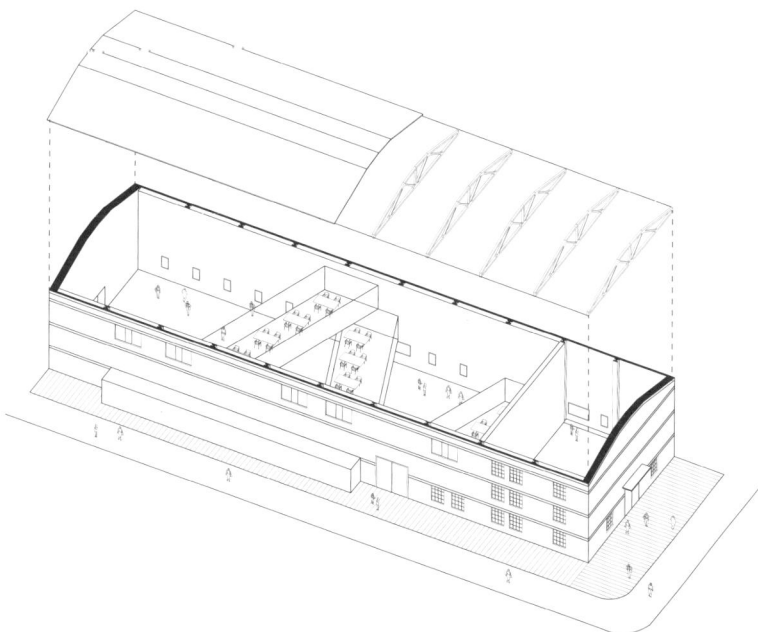

类型三

双坡单跨厂房

这类厂房的进深较小，提供适合办公、商业和餐饮的空间，也可以作为小型画廊使用。该类型的另一个优势是房屋单元之间紧靠在一起，提供了功能使用的灵活性，让创意企业之间可以实现相连或合并。

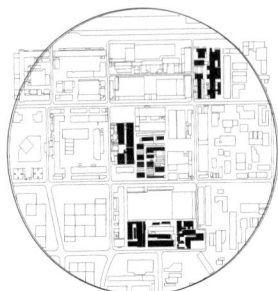

Type 3

Single-span factory buildings
with duo-pitched roof

With small building depth, they are suitable for office, business and catering, as well as small galleries and creative working. Another advantage of this type is that units are closely aligned, providing flexibility in use, in which creative enterprises can be connected or merged if needed.

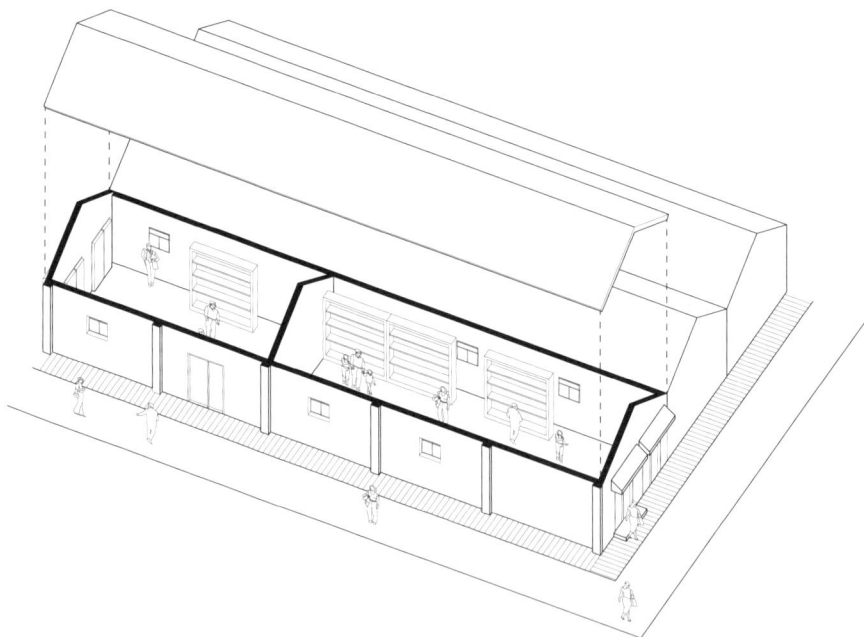

类型四

多层通用厂房

这也是一种很常见的建筑类型，它的外观特色不强，内部空间却提供了极高的功能改造的灵活度。研究的很多通用厂房底部楼层都被改造成服务旅游的商业设施，上面的楼层则可能为办公、酒店甚至是宿舍。

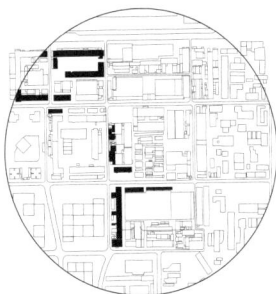

Type 4

Multi-storey factory buildings
for manufacturing

This is also a very common type of building in the 798 Art Zone. Buildings in this type have less visual identity, but they provides a great flexibility in reuse. Many ground floors have been converted into commercial facilities for tourism, with the upper floors likely to be converted into offices, hotels or even dormitories.

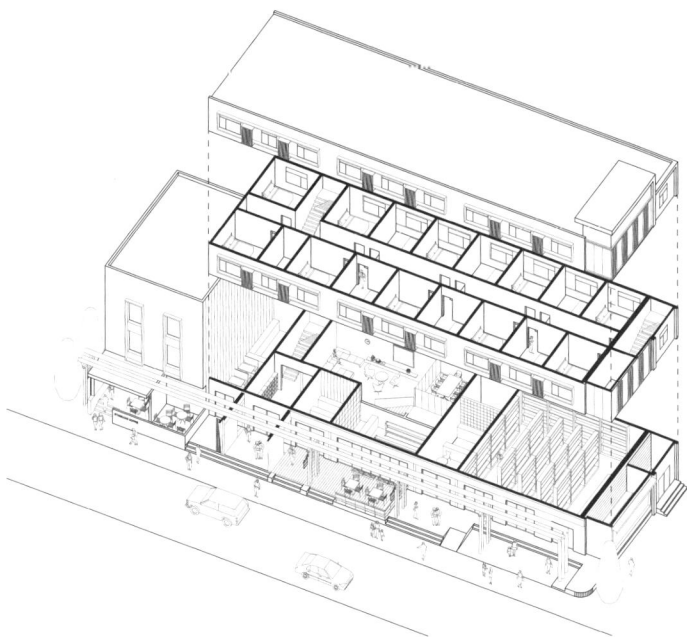

大芬村

5.5 | DA FEN VILLAGE

　　大芬村是中国城市化过程中一个特例。作为家装装饰的油画，在这里被制作和销售，并运往全世界各地。一个平凡无比的城中村，竟然能在全球的油画制作产业中占据一个位置，跃身成为产业重镇和旅游胜地，这不能不说是一个奇迹。产业的发展也给城中村带了如此巨大的变化，主要道路和中心广场不仅是村民日常休闲的空间，还成了画廊工作室争奇斗艳的所在。大街小巷里挂满了油画产品和半成品。透过橱窗，无数商家和画工正在忙碌着。在2008年金融风暴之后，新的油画品类和原创型艺术产业居然得到发展的机会，成为产业链中新的组成，这让人不得不赞叹产业发展的灵活性。从村民的角度，从务农为主转变为经营房屋租赁，甚至有的人自己干起了画廊生意，这是怎么样的一种身份转换啊！从这层意义上说，这就还不仅是一个单纯的经济现象，它的社会和文化意义也等待着我们进一步去发掘。

Da Fen Village is a special case in the process of urbanization in China. Oil paintings, as productized home decoration rather than as art, are made and sold here and shipped all over the world. It is a miracle that an ordinary village can occupy a position in the global oil painting industry and become an important industrial town and a tourist destination. The development of industry has also brought spatial changes to the village. The main roads and central square are not only the space for daily leisure of the villagers, but also become the place where the galleries and studios contend with each other. Oil paintings and semi–finished products are hung on sides of streets and alleys. Through the display windows of galleries, countless merchants and painters are busy. In 2008, after the financial crisis, the new oil painting categories and the original art industry has surprisingly seized the opportunity to become a new component of the industrial chain. During the development, the business of villagers shifted from farming to operating house rental, and some villagers even set up their own gallery business. This is such a kind of identity conversion! In this sense, the happening in Da Fen Village is not only a simple economic phenomenon, its social and cultural significance is also awaiting further exploration.

城市空间肌理图
Figure ground plan map

卫星图
Satellite map

路网与广场
Roads and plazas

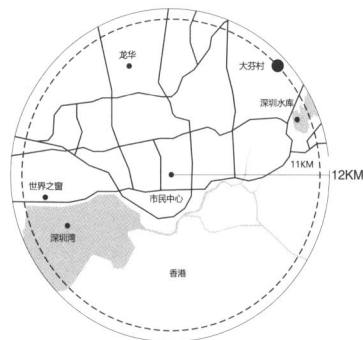

区位图
Location

和其他城中村一样，大芬村的城市结构也几乎是完全以自我为中心，与外界的空间体系关联不大。村中几条主要的空间轴线是沿着主要机动车道路展开的，这里是村民活动和画廊招揽顾客的主要场所，不乏活力，但是环境不佳，改善的空间还很大。有意思的是空间主轴一直处在生长过程中，并向村子的内部进一步渗透，在这个趋势下，目前有一些村内的小街巷也得到了一定更新。这里还有一些特殊的节点，如大芬村美术馆。

Like other urban villages, the urban structure of Dafen Village is almost entirely self-centered and has little to do with the spatial system of the surrounding areas. Several main spatial axes in the village are launched along the main vehicle roads, and these are the main place for villagers' activities and galleries to find customers. There is no lack of vitality, but the environment quality is not ideal, and there is still a lot of room for improvement. Interestingly, these main spatial axes have been, in the growth process, further penetrating into the village. With this trend, some small streets and alleys in the village have also been updated to a certain extent. Some urban nodes exist here and there, such as the plaza before the Da Fen Village Art Museum.

大芬村的产业结构是非常清晰的。主导产业为绘画工作室、画廊和绘画原料相关行业，占据较大的比例。以主导产业关联，一些艺术教育、艺术品商店和博物馆产业也比较兴盛。村民大多仍旧住在这里，服务他们的学校和商业设施仍然留存，这让大芬村的社区氛围浓厚。大芬村里面向旅游的商业设施不多，这在几个案例中是比较独特的，其原因很可能是商业的利润不及画廊。

Da Fen Village's industrial structure is clear, whose dominating category contains painting workshops, galleries and manufacturing of raw materials related aforementioned industries, which accounts for a large proportion. Linked to the leading industries, some art education, art shops and museums are also thriving. Most of the villagers still live here, and the schools and commercial facilities that serve them are still there, which gives Da Fen Village a strong community atmosphere. The relatively low proportion of commercial facilities for tourism in the village is unique among the other cases presented in this book, probably because their commercial profits are less than those of galleries here.

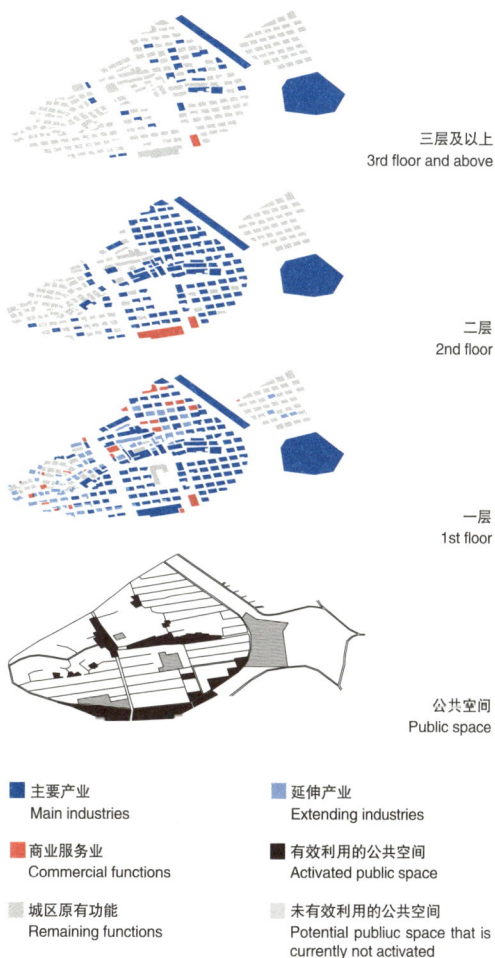

三层及以上
3rd floor and above

二层
2nd floor

一层
1st floor

公共空间
Public space

■ 主要产业
Main industries

■ 商业服务业
Commercial functions

▨ 城区原有功能
Remaining functions

■ 延伸产业
Extending industries

■ 有效利用的公共空间
Activated public space

▨ 未有效利用的公共空间
Potential publiuc space that is currently not activated

功能分类 Category of functions	具体功能及企业 Function description and enterprise	比例 proportion
保留的城区原功能 Remaining functions	住宅及附属设施 Housing and affiliated facilities	33%
商业服务业 Commercial functions	餐饮店（111家）Gastronomy (111)	2%
	游客中心（2家）Tourist centre (2)	
	零售店（5家）Retails (5)	
	打印，网吧，按摩（5家） Print shop, Internet cafe, Massage (5)	
主要产业 Main industries	工作室（87家）Studio (87)	39%
	画廊（210家）Gallery (210)	
	木工，画框，画布等（75家） Carpentry, Picture frames, Canvas, etc. (75)	
延伸产业 Extending industries	美术用品商店（9家）Art supplies shop (9)	26%
	书店（5家）Bookstore (5)	
	艺术学校（2家）Art school (2)	
	博物馆（1家）Museum (1)	

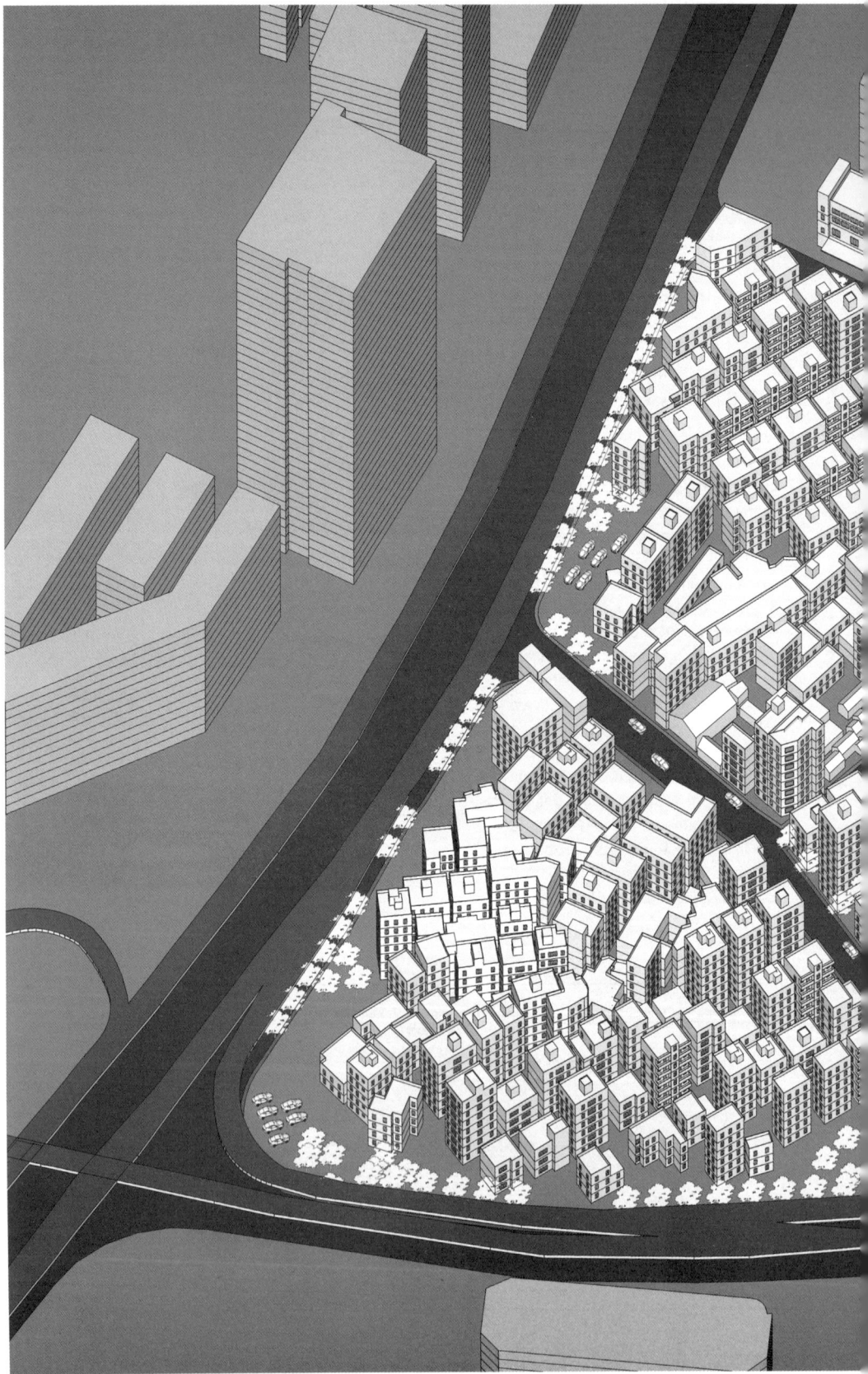

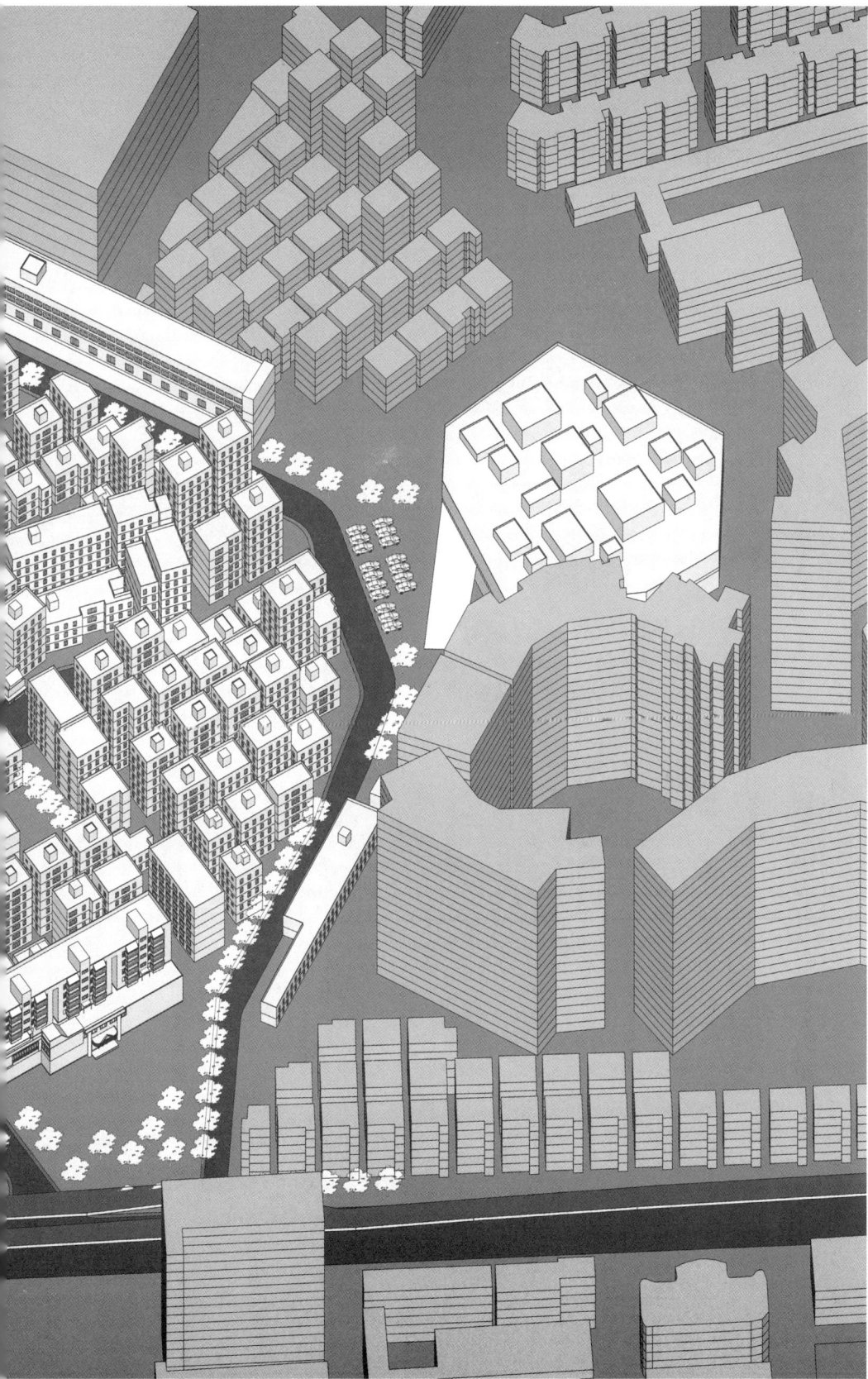

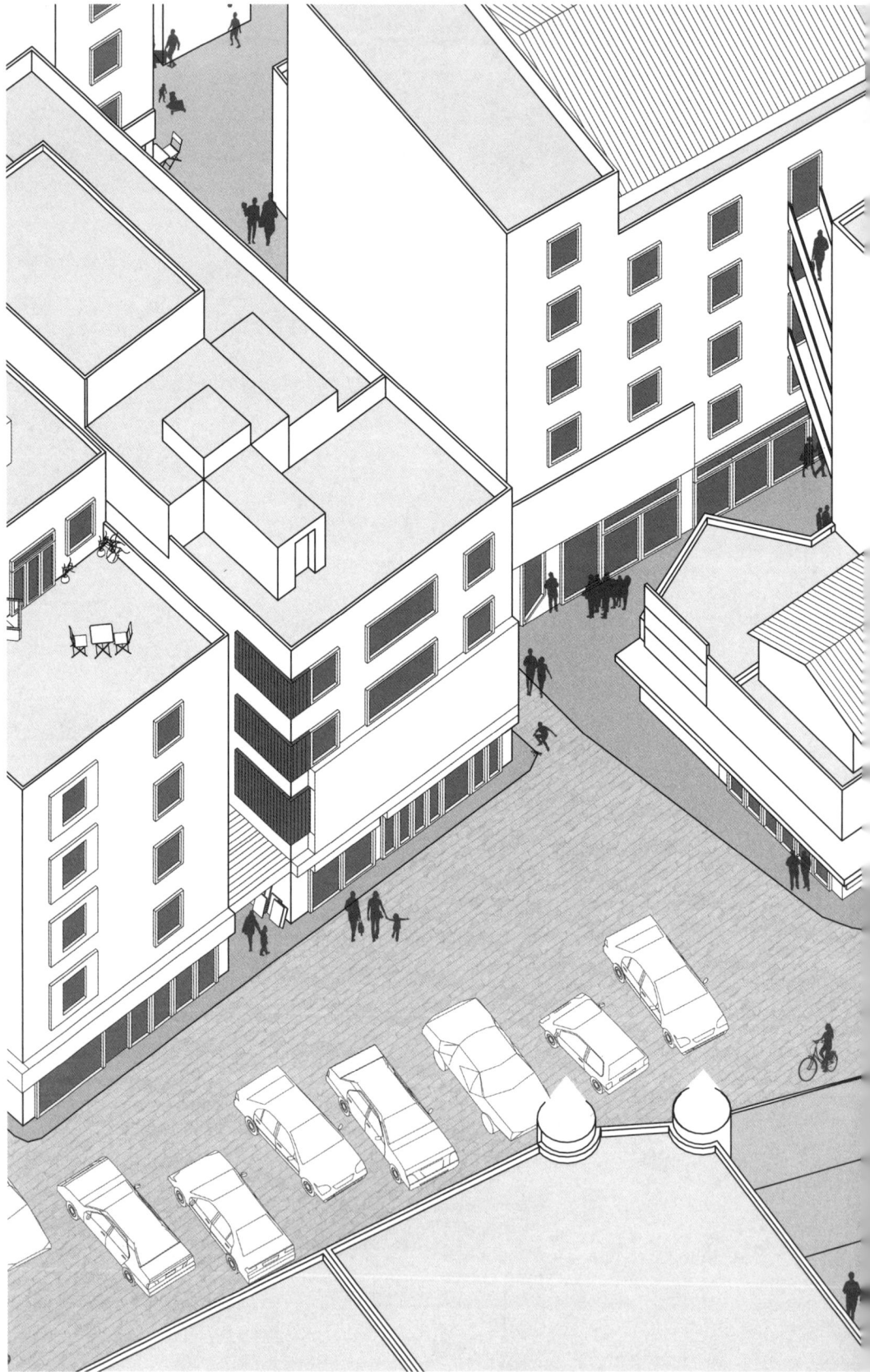

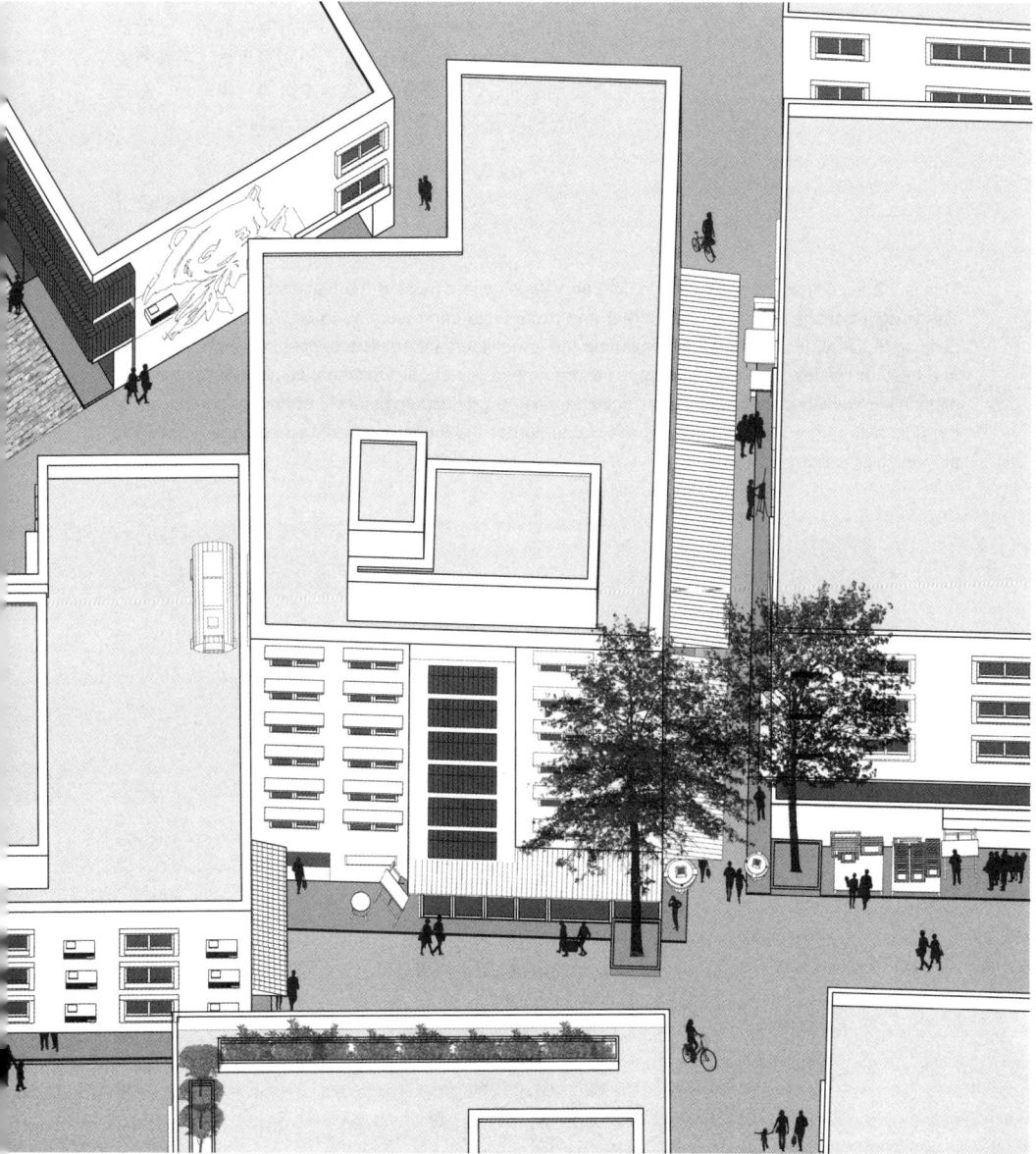

类型一

多层村民住宅

　　这是大芬村最为典型的空间单元，这里大部分的房屋都属于这种类型。沿着主街和广场的村民住宅，功能都被重新利用以适应油画产业的需求，例如下面的楼层改造为画廊，面向商业街开门和进行橱窗展示。上面的楼层则有多样的功能组合可能，如油画作坊、画具相关产品的车间、画工宿舍等。远离主街的空间单元，则产业的比例逐渐下降，村民自住的比例较高。

Type 1

Multi-storey villager houses

This is the most typical building unit in Da Fen Village, where most of the houses belong to this type. The villager houses along the main street and plaza have been adapted to new use related with oil painting industry, for example, by converting the lower floors into galleries, opening doors to streets and making window displays. The upper floors have a variety of functional combinations, such as oil painting workshops, workshops for painting-related products, painters' dormitory and so on. In the unis away from the main streets, the proportion of the industry gradually decreases, while the proportion of residence increases.

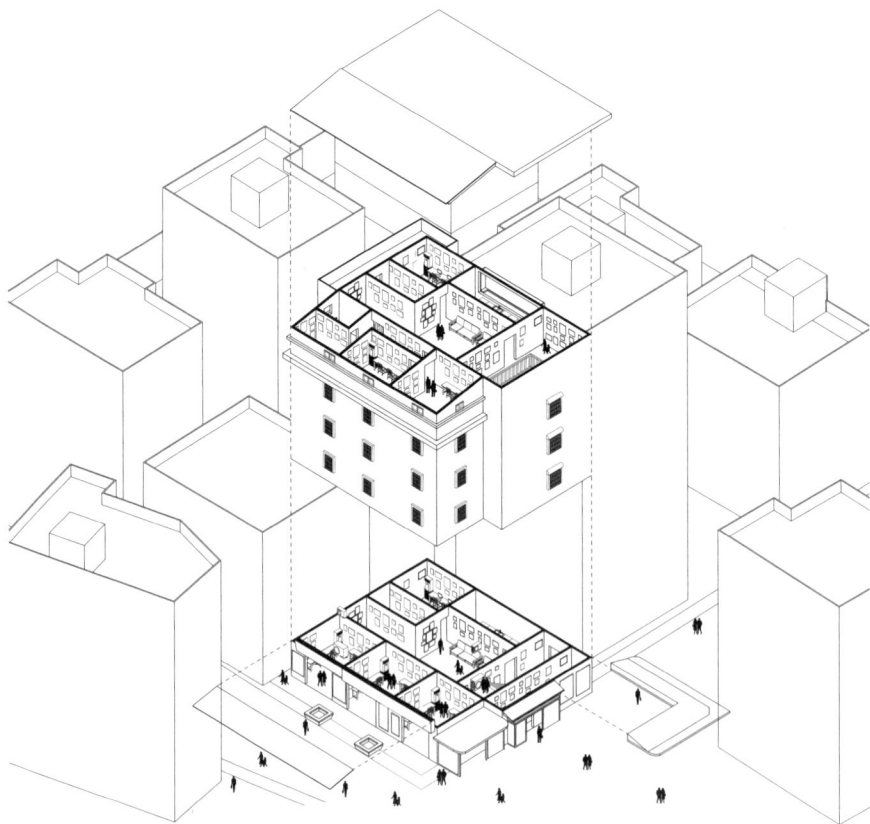

类型二

尺寸稍大的多层建筑

此类建筑尺寸比村民住宅略大，原来可能是宿舍、小型办公楼或小型多层厂房。他们的功能改造模式和村民住宅类似。

Type 2

Multi–storey buildings

They are slightly larger than the villager houses, which uses to be dormitories, small office buildings or small factory buildings. Their functional transformation model is similar to that of villager houses.

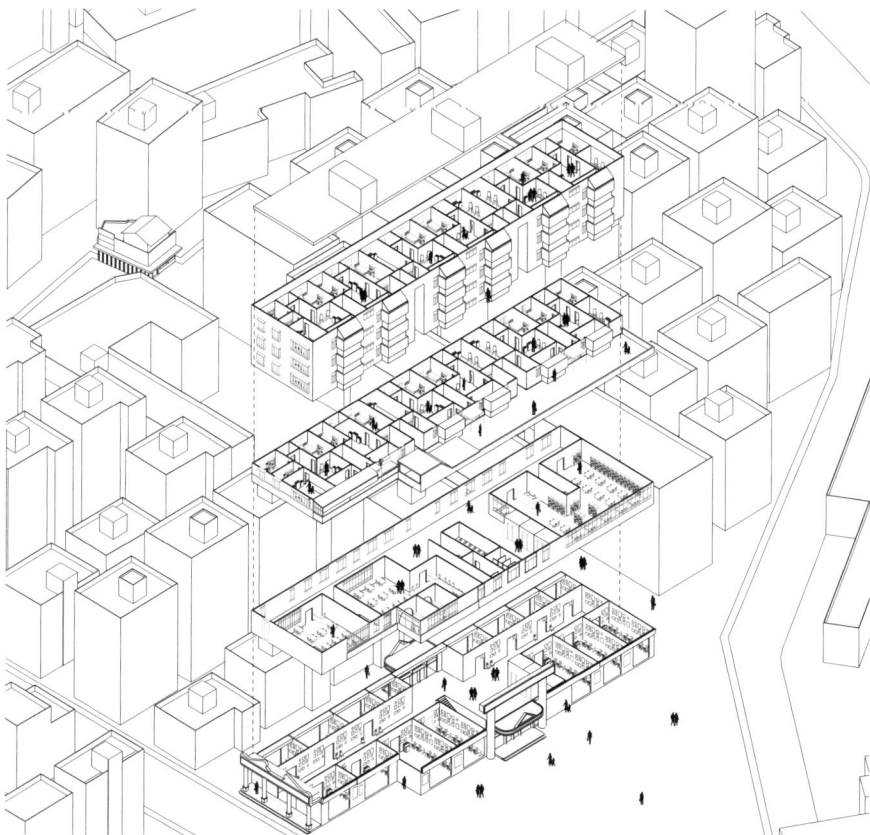

特殊类型一

极长的建筑

这是位于大芬村东北角的一栋特别狭长的建筑，某种程度上，它就像一段城墙，定义着大芬村的这段边界，并给经过的人留下深刻的印象。目前它只是一些底层商店和上部的混杂的功能（库房、宿舍、旅店和小型展厅）。让我们想象一下，未来，它会被变成一个拥有最长的T台的时装秀场，还是一个狭长的艺术品市场？

Special type 1

Extremely long building

This is a narrow and long building in the northeast corner of Da Fen Village. In a way, it is like a section of the city wall, defining this part of village boundary, and making a deep impression to the people who pass by. At the moment it is just a mix of shops on the ground floor and other functions on the upper ones (storehouses, dormitories, hotels and small showrooms). Let's imagine, if in the future it can be turned into a fashion show hall with the longest catwalk? Or a unique art market?

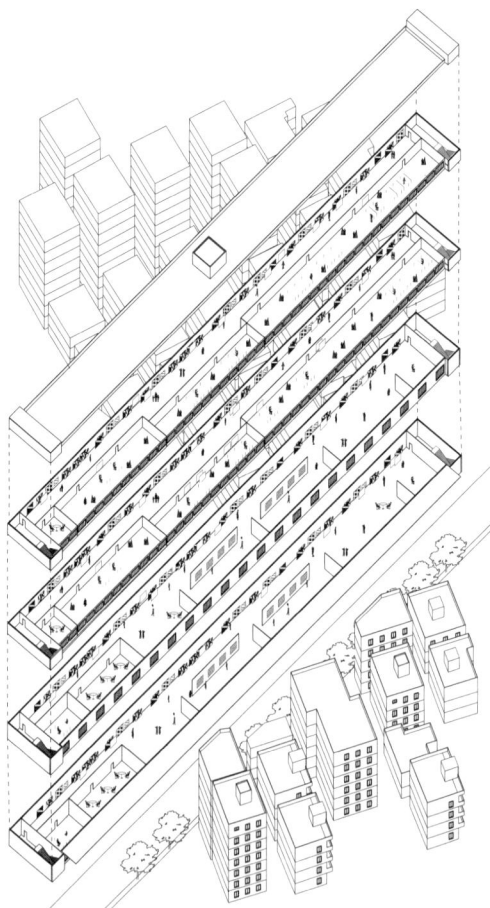

特殊类型二

大芬美术馆

在一大群小体量建筑之中，大芬美术馆是很独特的。它位于 T 字形主街交汇处为，在博物馆巨大的身影前，留下同样巨大的广场。它太整洁，整洁得好像它并不属于这里。这块场地什么时候能够从空旷的美术馆的前广场，变成大芬村真正的生活性广场呢？还有，美术馆的存在，对新兴的原创性绘画产业，能不能发挥更大的作用呢？

Special type 2

Da fen art museum

Located at the junction of T-shaped main streets, it is unique among a large group of small buildings. In front of the huge figure of the museum, it leaves the huge square, which is maybe too neat and tidy as if it doesn't belong here. Is it able to be changed from the "front museum square" to the real living square of Da Fen Village? Furthermore, can the museum play a greater role in the emerging original oil painting industries? Also, can it play a great role for the emerging original art industry?

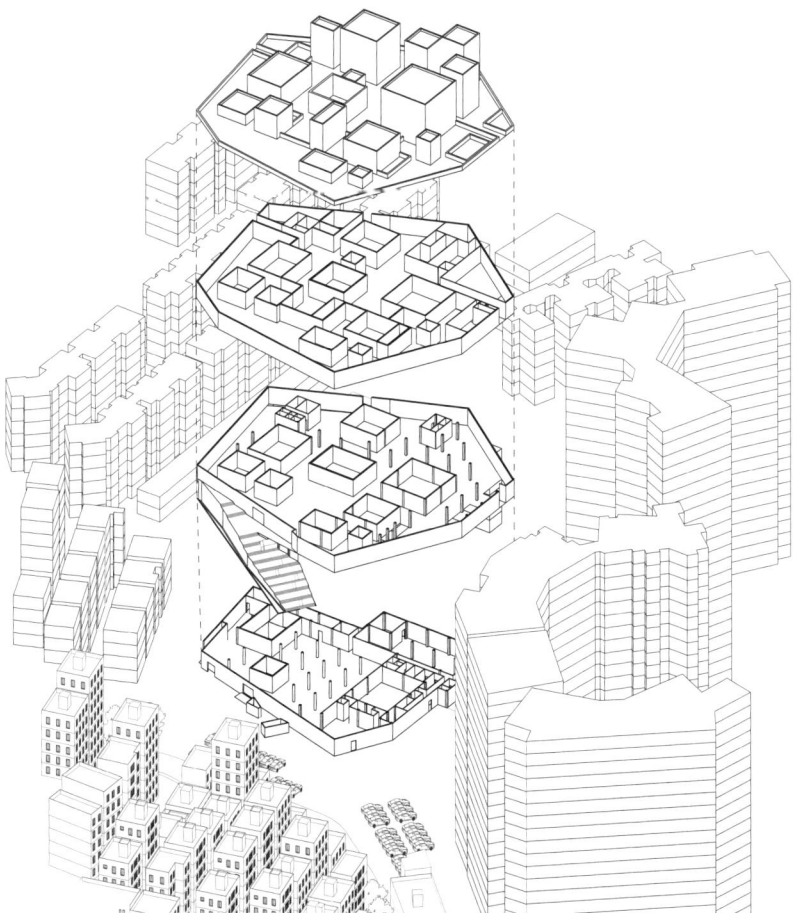

田子坊

5.6 | TIAN ZI FANG

　　田子坊在国际上享有盛名。它向全世界展示了里弄建筑这种上海独特的居住建筑形式，更是展示了一种城市发展可能性。没有大拆大建，这里的建筑大多都还是原来的样子，只是建筑里的内容和城区的氛围发生了那么大的变化。宁静的宅间通道变成了熙熙攘攘的商业街道；居民逐渐迁出，取代的是商家和企业，原本破旧的建筑被翻新和改造。如果你在 2010 年以前来到这里，你会发现还有那么多居民住在这里（当然，当时有很多住宅已经被转租给了画廊和工作室）。年轻的艺术家和居民出没于弄堂之间，他们之间存在多么有趣的一种关系啊！业主和租户之间合作和摩擦无时无刻不在发生，他们曾经一起组织过抗议开发商"强拆"的运动，也常因为日常生活中一些问题纠纷不断。随着居民的搬出，以及艺术家们因为支付不起日渐增高的房租也渐渐搬出，今天这些微妙的关系都不太常见了。今天的田子坊更像是一个特色的旅游区，某种程度上创意的氛围已经比较弱了。尽管如此，它仍是上海老城区最具吸引力的一个去处。

Tian Zi Fang enjoys a great reputation in the world. It shows the world the unique residential architecture type of historic Shanghai, as well as an unexpected possibility of urban redevelopment. Without demolition and mass new construction, most of the buildings here remain in its original state, with the contents of the buildings and the atmosphere of the urban area having changed intensively. The former quiet passageways between housing units have been turned into bustling commercial streets, while residents have gradually moved out to give more place to businesses. Formerly dilapidated buildings have been renewed and renovated. If you came here before 2010, you'd find so many original residents still living here . What an interesting relationship there is between the artists and the residents who pass by the alleyways! Cooperation and friction between owners and tenants take place all the time, when they have organized protests against developers "demolition" campaign, or when they came into problems in daily life disputes. As residents move out, and artists move out because they cannot afford the rising rents, these delicate relationships are less common today. Today's Tian Zi Fang is more like a characteristic tourist area, to a certain extent, the creative atmosphere has been relatively weakened. Nevertheless，it still is one of the most attractive places in the old city of Shanghai.

蛋卷鞋®
UICHE SHOE

城市空间肌理图
Figure ground plan map

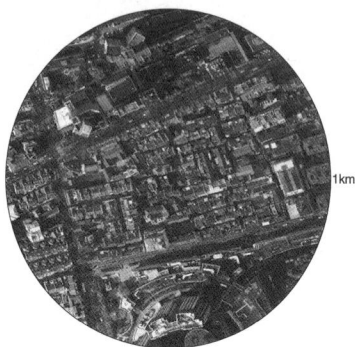

卫星图
Satellite map

路网与广场
Roads and plazas

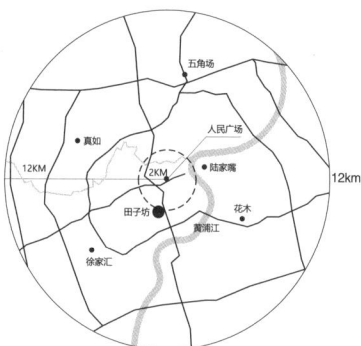

区位图
Location

　　密集的网络状街巷是田子坊空间结构的支配性要素。这些街巷原来是住宅区的半公共通道，现在则被纳入城市公共空间的系统，它和南边的泰康路连在一起，形成若干个主要出入口。随着田子坊进一步发展，这张网络还在进一步朝着外扩张，打通相邻住宅区和工业地块的街巷通道，把它们也一起拉入到城市更新的进程中来。目前朝着北部的建国路和西部的瑞金路，新的出入口已经形成，很多人流也可以从那里穿过原本封闭的城区，进入田子坊。

Dense network shaped by streets and alleys are the dominant element of the spatial structure of Tian Zi Fang. Originally semi-public passageways for residential areas, these alleys are now integrated into the urban public space system of city, which connects with Taikang Road in the south and forms a number of main entrances there. As Tian Zi Fang further develops, the network is also expanding further outwards, opening alleyways to neighboring residential and industrial plots, and thus bring them together into the process of urban regeneration. At present, toward Jianguo Road in the north and the Second Ruijin Road in the west, new entrances have been formed, by which people can now pass through the originally closed urban areas to enter Tian Zi Fang.

作为主导产业，文化艺术产业引领了区域的第一轮发展。近年来延伸产业发展很快，尤其是服饰商店和纪念品商店，与之同样发展迅速的是餐饮业。目前这些主要针对游客的产业已经占到很高的比例，让主导产业倒显得不那么突出了。另一方面，田子坊一直就是一个居住社区，仍然有比较多的居民居住在这里。城区原有功能在田子坊中心区域比较少，越靠近边缘越多，直至完全融合在周边的居住社区中。

As main industries, culture and art industry have led the first round of regeneration in Tian Zi Fang. In recent years, various extending industries have developed rapidly, especially in clothing stores and souvenir shops, while gastronomies has also developed rapidly. These industries, which are mainly oriented to tourists, now account for a high proportion, making the leading industries seem less prominent. On the other hand, Tian Zi Fong has always been a residential community, and there are still many residents living there. These remaining functions are less in the central area of Tian Zi Fang, whose proportion increases when it is close to district's edge, until the completely integrating in the surrounding residential communities.

二层及以上
2nd floor and above

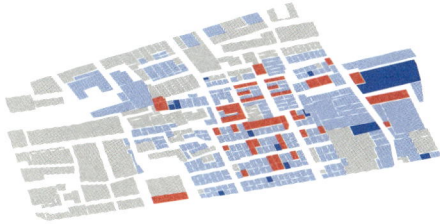

一层
1st floor

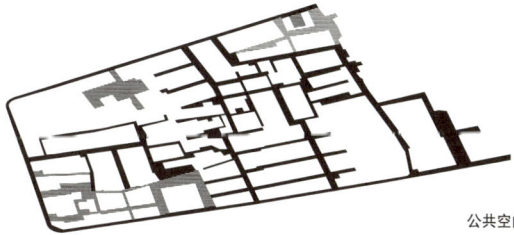

公共空间
Public space

主要产业
Main industries

商业服务业
Commercial functions

城区原有功能
Remaining functions

延伸产业
Extending industries

有效利用的公共空间
Activated public space

未有效利用的公共空间
Potential publiuc space that is currently not activated

功能分类 Category of functions	具体功能及企业 Function description and enterprise	比例 proportion
保留的城区原功能 Remaining functions	住宅及附属设施 Housing and affiliated facilities	58%
商业服务业 Commercial functions	餐饮店（47 家）Gastronomy (47)	5%
主要产业 Main industries	艺术家工作室（21 家）Artist studio（21） 画廊（36 家）Gallery（36） 文化产业（55 家）Culture industry（55）	23%
延伸产业 Extending industries	工艺品设计展示（62 家） Display of handicraft design（62） 服饰设计（91 家）Dress design（91） 旅游纪念品（37 家）Souvenir shop（37）	14%

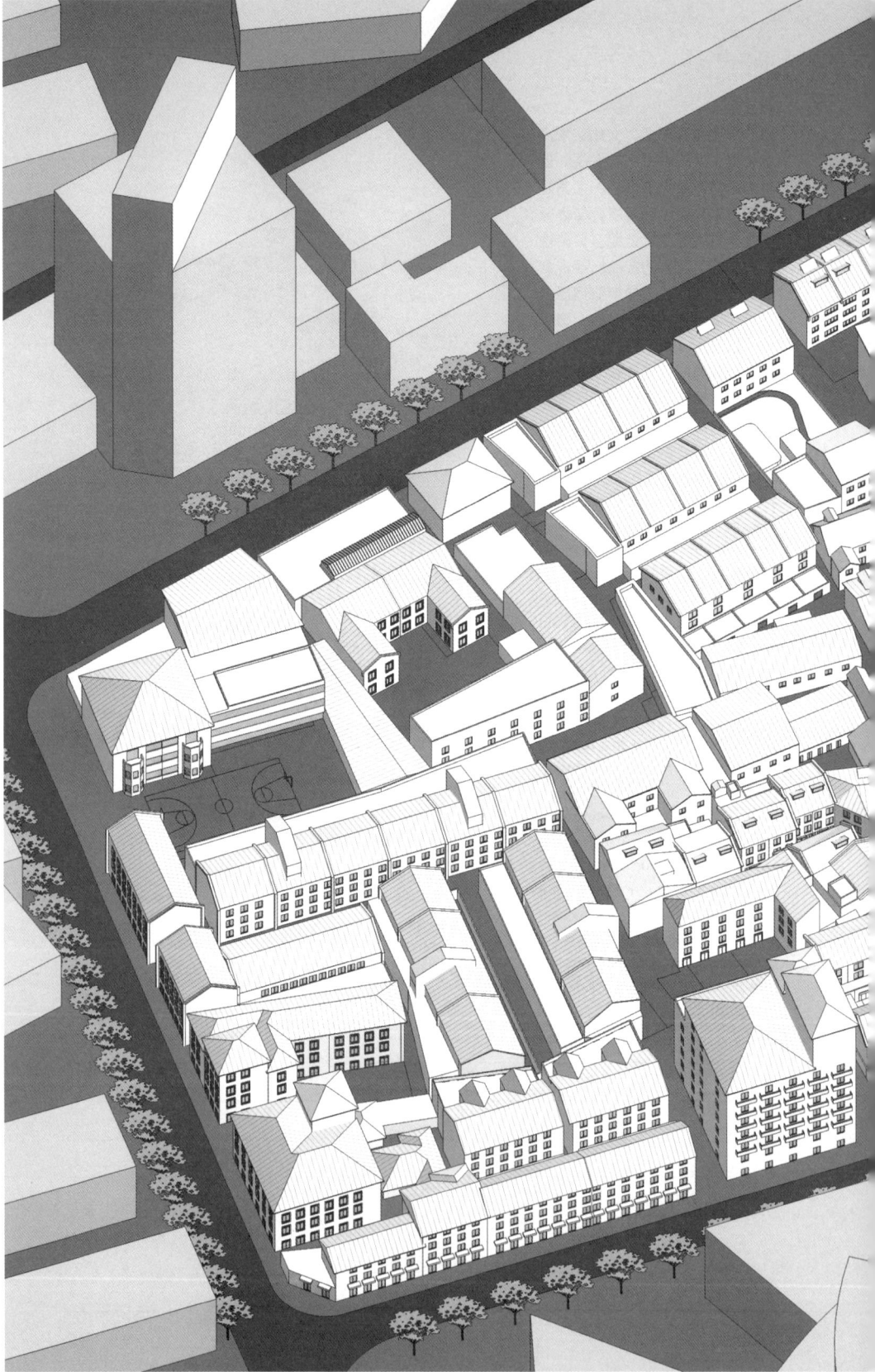

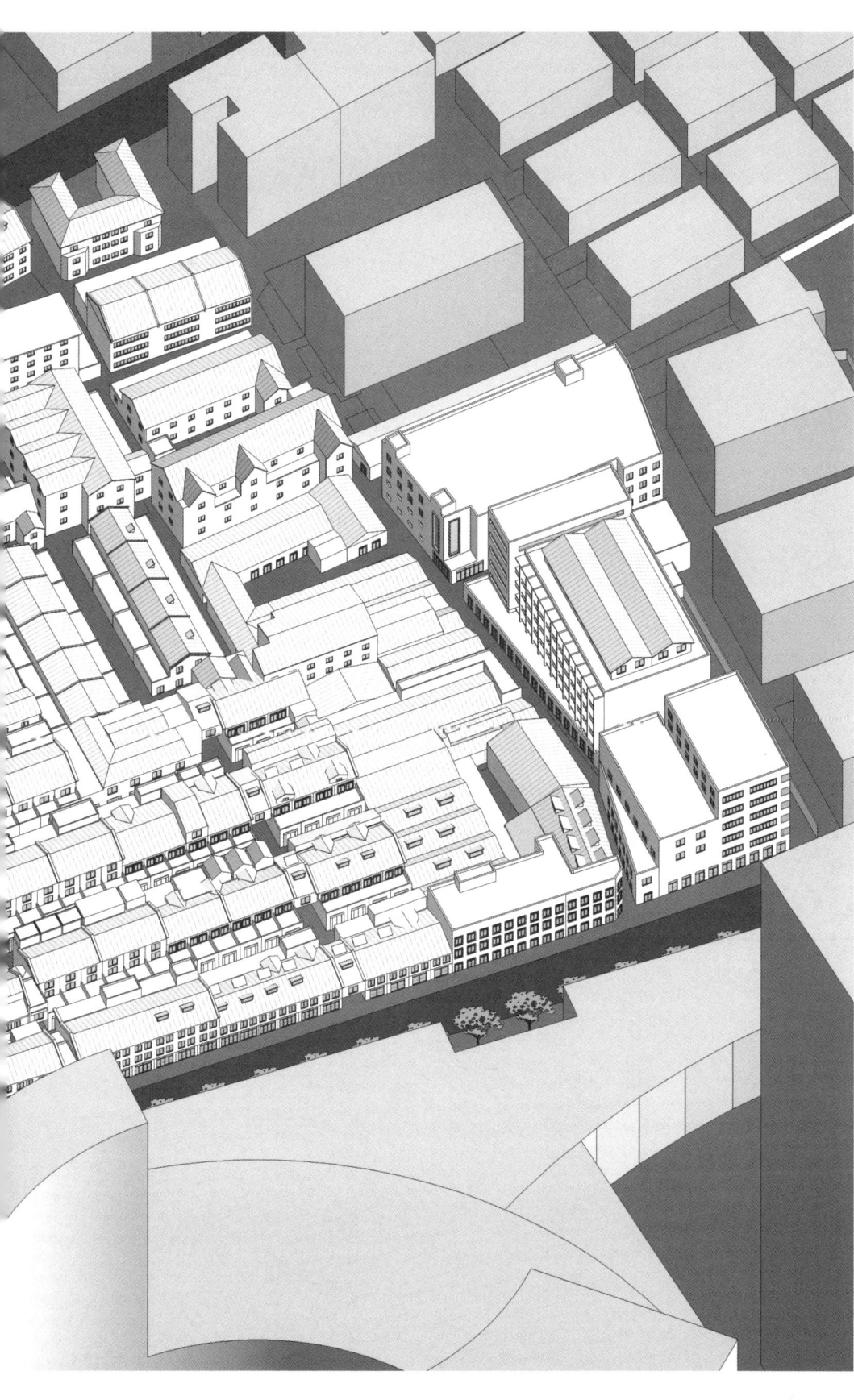

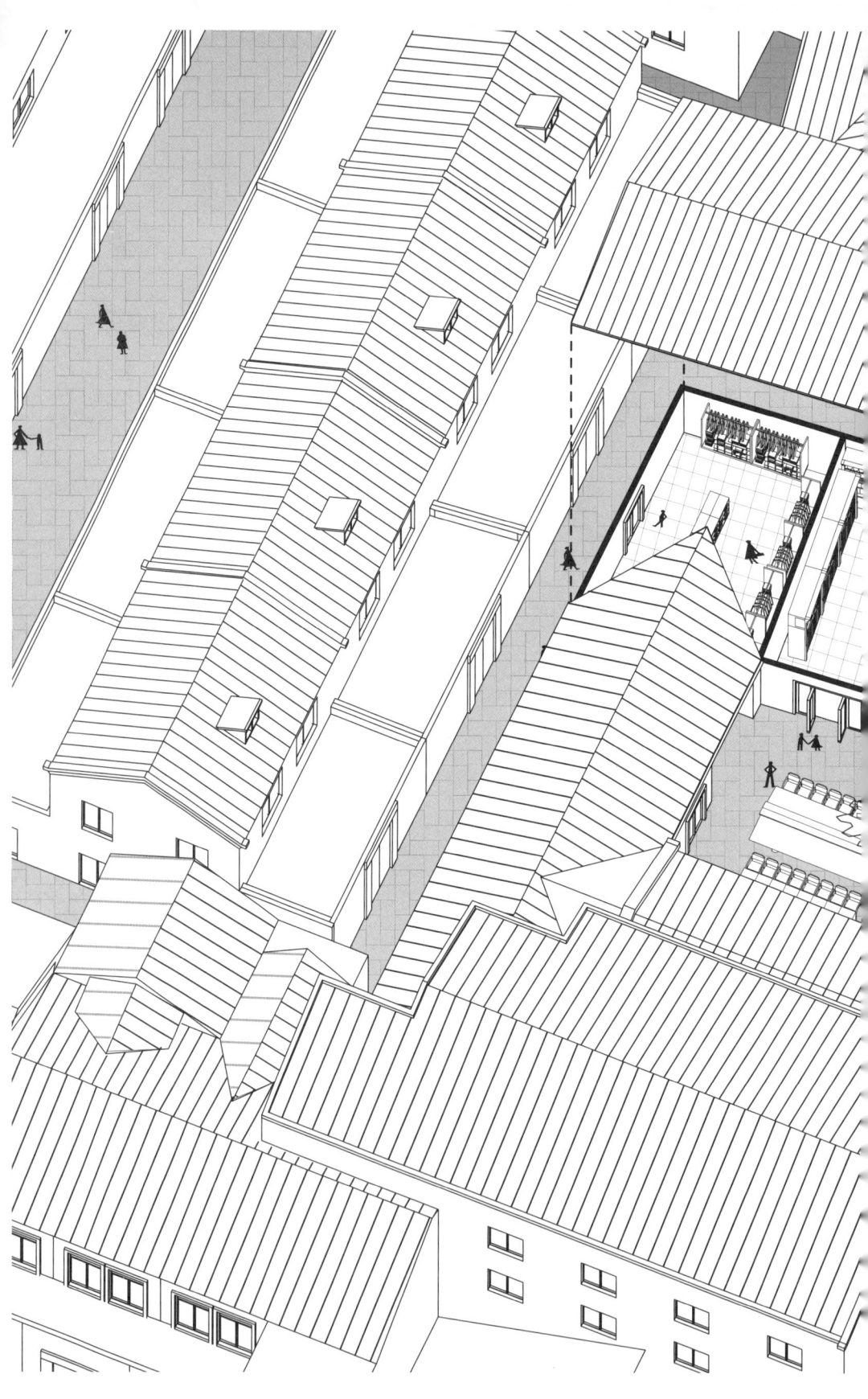

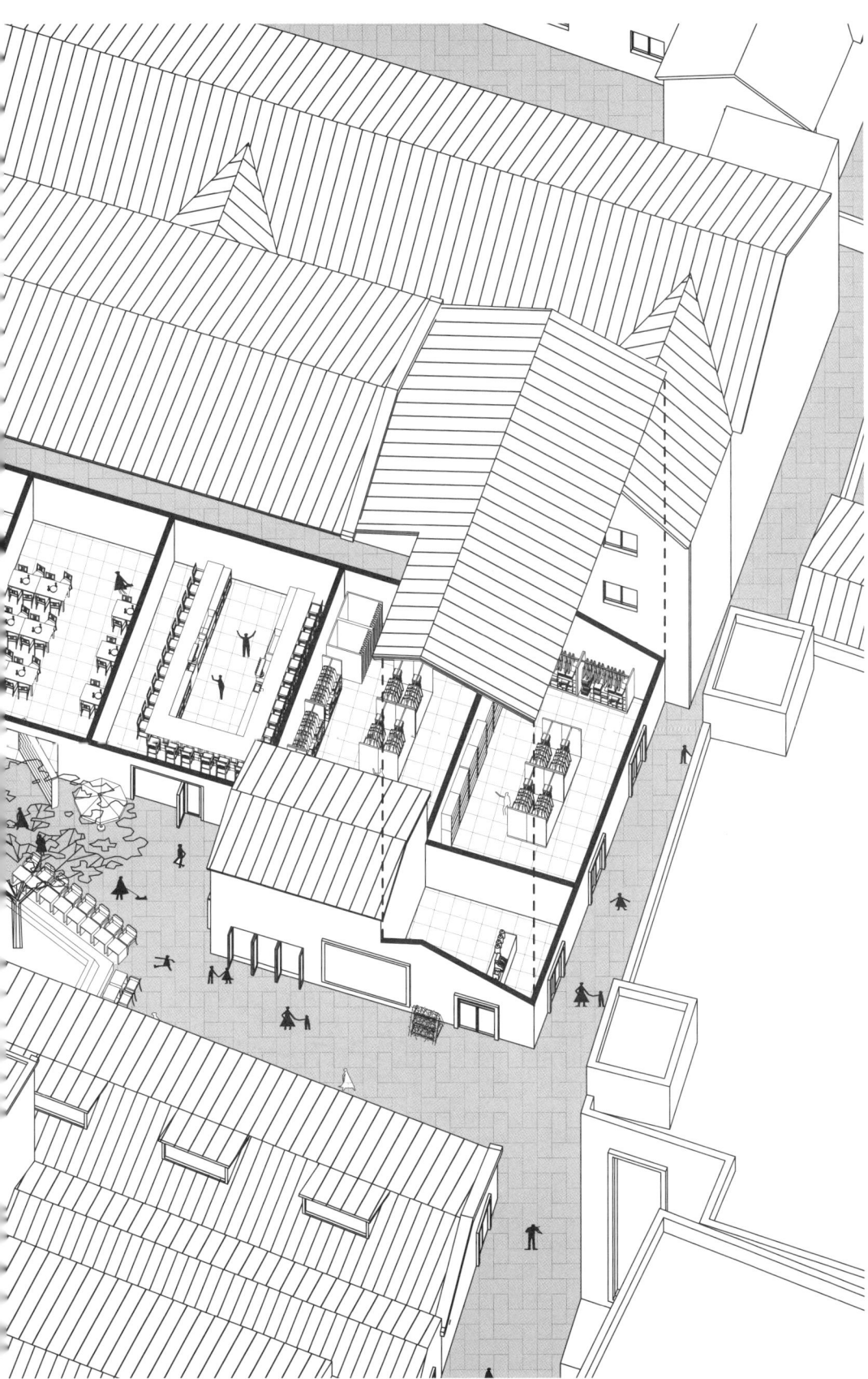

类型一

单开间里弄建筑

　　这种联排式建筑是田子坊最典型的空间形式。产业或者商户入驻后，首先会将原来的入户门（石库门）的门板换成玻璃橱窗。之后，他们会用玻璃覆盖入户庭院，把这些空间巧妙地做成商品（产品）陈示区。这样，带有历史风格的石库门外框，加上现代化的玻璃橱窗和玻璃天窗，往往成为展示商品的奇异而富有魅力的空间。建筑外的空间，会被精心装饰成一个特殊的领域，吸引游客驻足。建筑的内部改造则千变万化但往往会保留一部分原有建筑的特色。

Type 1

Standard single-bay Li Long housing unit

This is the most typical space form of Tian Zi Fang. After the enterprises enter the buildings, they will first replace the original doors (so called "Shi Ku Men") with glass and make them into display windows. After that, they cover the small courtyard with glass and steel structure, cleverly turning the space into a display area for goods . In this way, with Shi Ku Men's outer frame with strong historic style, coupled with modern glass windows and glass roof, a curious and attractive display space comes into being. Spaces outside buildings will be meticulously decorated into a special area, attracting tourists to stay. The internal transformation of the building is ever-changing but often retains some of the characteristics of the original buildings.

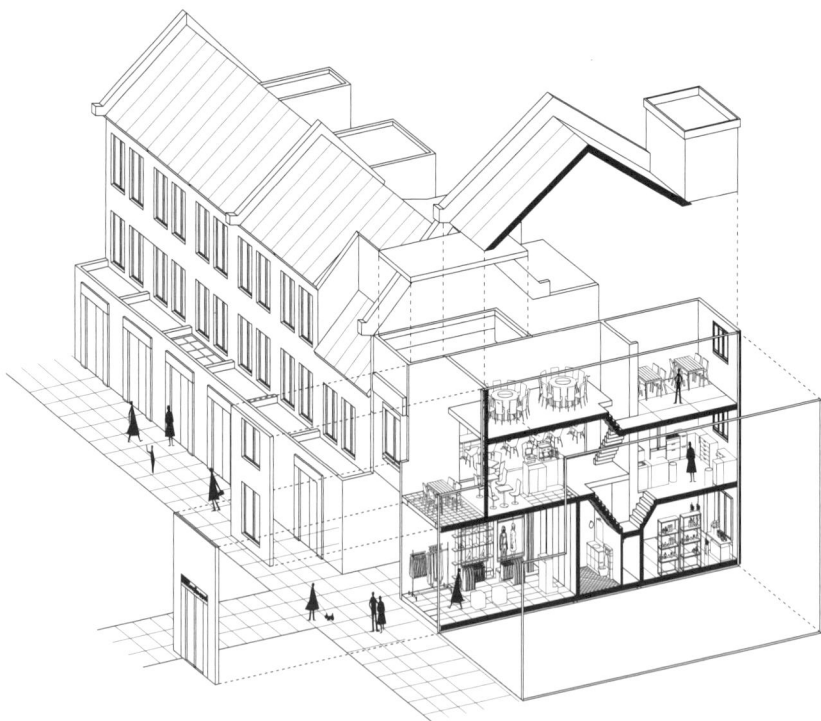

类型二

多开间里弄建筑

有个别的里弄建筑单体，有较大的前院或后院，或者有一些特殊的建筑布局。对它们的改造方式则非常多样。很多院子同样会被覆盖，面向外部街巷的橱窗也同样非常重要。

Type 2

Multi-bay Li Long housing unit

Some residential building units have larger front-yards or backyards, or special architectural layouts. Their spaces are reorganized it in a variety of ways. Many courtyards are also be covered, and display windows facing the exterior alleys are also designed carefully.

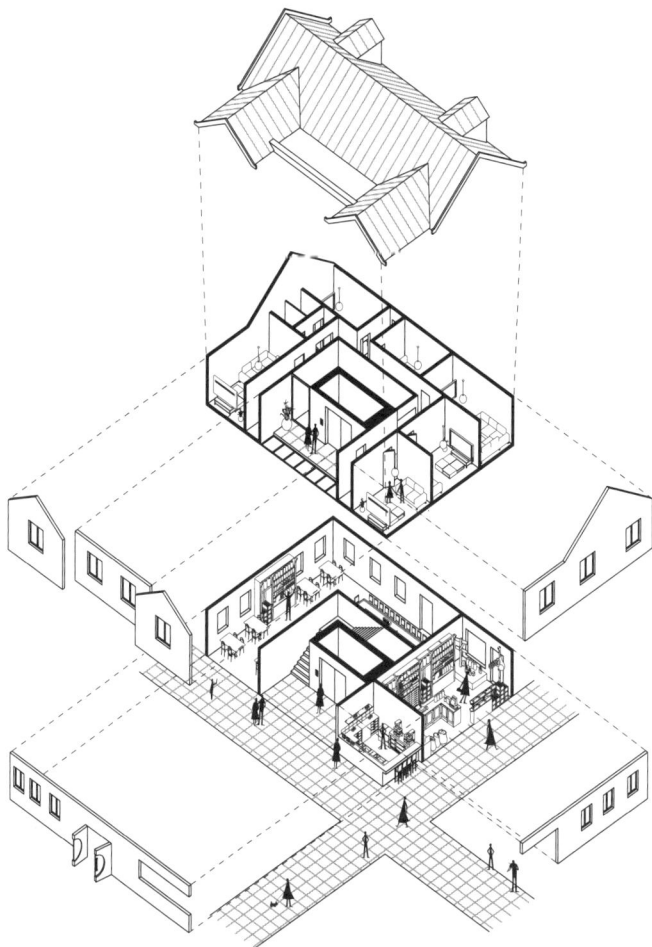

类型三

多层建筑

这是一些多层平屋顶建筑，它们以前曾经是工厂车间或附属办公。这样的建筑，在田子坊的内部有一些，外围（沿着泰康路和瑞金二路）也有一些。它们不具有前面所述的里弄建筑的商业价值和历史特色，但可能因为不太高的租金成为创意产业喜欢的房屋形式。建筑内部的空房间可以被灵活划分以适应不同产业的需求。底层如果沿街的话，则很可能还是被改造成商铺的功能。

Type 3

Multi-storey buildings

These were formerly some industrial buildings or affiliated office buildings. They are located either within the territory of Tian Zi Fang or in its border (along Taikang Road and Ruijin Road). They do not have the commercial value of the Li Long housing units mentioned above, but just because of that, they may become the preferred form of housing for the creative industry for relatively low rents. The empty rooms inside the building can be flexibly partitioned to suit the needs of different enterprises. If the ground floor is along the street, it is likely to be transformed into commercial functions.

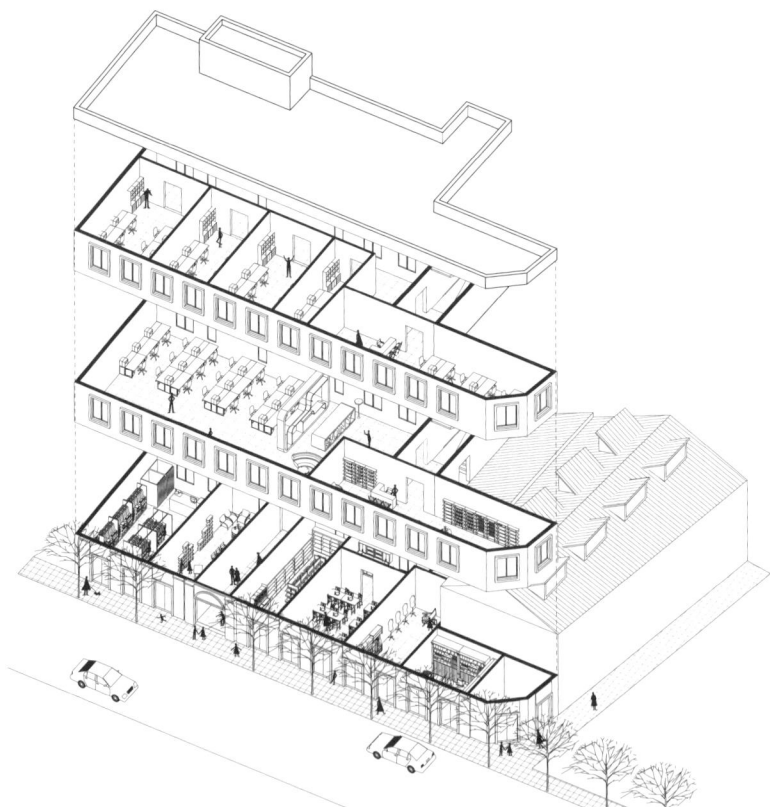

类型四

单层双坡仓库建筑

田子坊内部有几处这样的房屋形式。当室内的吊顶被打开，原始的屋架被暴露和展示的时候，空间的美会震撼每一位参观者。在田子坊发展的早期，这些房屋首先被一些知名艺术家看中，成为他们的工作室和展示空间。今天，这些空间仍然大部分用于服务艺术功能，如名人纪念馆、博物馆和画廊等。

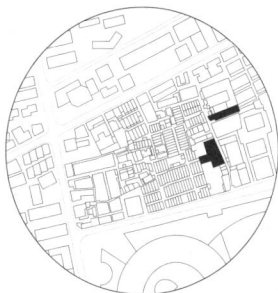

Type 4

Single-storey warehouse building with duo-pitched roof

There are several such buildings in Tian Zi Fang. When the ceiling is opened and the original roof truss is exposed and displayed, the beauty of the space will shock every visitor. In the early days of Tian Zi Fang's development, these warehouses were first valued by some famous artists as their studios and exhibition spaces. Today, most of these empty rooms still serve artistic functions, such as the halls of fame, museums and galleries.

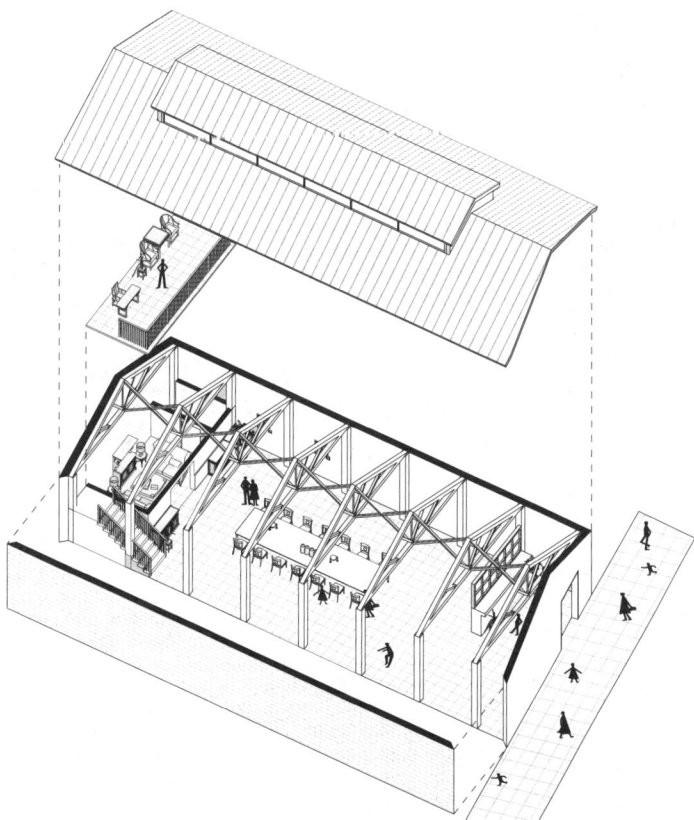

6 城市设计在自发更新中的作用
THE ROLE OF URBAN DESIGN
IN BOTTOM-UP URBAN REGENERATION

6.1 "自上而下"与"自下而上"的关系
BETWEEN "TOP-DOWN" AND "BOTTOM-UP"

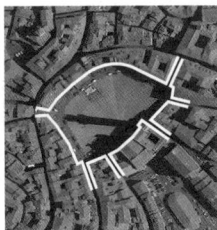

图 35 坎波广场、圣马可广场和
人民广场里城市建筑共同遵守的
界面

Fig.35 Regulated urban interface
that buildings shall all follow, in
Piazza di Campo, Piazza San
Marco and Piazza Popolo

创意社区的出现，已经向我们表明，可持续城区发展的动力可以来自每一个个体，通过他们的需求和相互作用，让城市的物质空间得以更新发展。我们也已经证实，在这些城区的发展中，"自上而下"的空间导控手段，起着两面性的作用。一方面，固化的、对动态发生趋势不敏感，甚至是粗暴的规划和城市管理是这些城区发展公敌。当我们回想起798艺术区发展初期的那些粗暴的对抗和田子坊居民自发组织的反对官方规划的游行时，仍然可以感觉到那些对抗的惨烈程度。而大芬村和曾厝垵这样的案例，如果不是他们在"城中村"这样一个城市法规的避难地带，又怎么可能发展出如此庞大的创意集群。另一方面，难道"自上而下"可以完全缺位吗？没有对滨水岸线的拓宽和整治，沙坡尾的滨水空间就永远只是一个"好看"的所在，他不会像今天那样成为一个滨水场所，滨水一侧的产业也不可能再上台阶。同样，作为田子坊南边界和主要"视口"的泰康路，如果没有对道路使用的治理以及对服务设施的重新规划布局，今天仍是会嘈杂不堪的"马路菜场"，田子坊也不会是今天的样子。田子坊从当初的"边缘式"、"地下式"的发展，迎来全面的繁荣，还要感谢2008年全区被划入"创意产业园"，从此所有的改变住宅功能的行为被合法化，商家可以合法经营，受到保护。上面这些举措，不管是"空间"的还是"政策"的，都无疑起到了巨大的作用，而它们都不是仅仅依靠"自上而下"的途径实现的。

那么"自上而下"和"自下而上"的关系到底是什么？二者是不是有进行某种"合作"的可能性？或者，让我们把这个问题更进一步简化，那就是当我们运用"自上而下"的手段去尝试导控一个城区的发展时，它的边界在哪里？回顾我们在导论部分提出的愿景——将创意城区作为城市发展的一种具有启发意义的特殊现象——那么，这个问题可以并不特指创意社区的发展问题，而是对中国城市的发问。中国城市如何不再是单方面的"自上而下"的空间指令的产物，而是给予城市的利益相关者更多创造空间的权力？这样的话，城市是不是也可以以此获益，变得更具特色和活力，不可预期却令人惊喜呢？

我们在历史上可以找到那么多精彩绝伦的城市案例，它们告诉我们"自上而下"的恰当边界在哪里，而边界之外的地方，则能够让经济、产业和社会去通过一个个利益相关者发挥它们的作用。当然，这边界需要多么巧妙地去界定啊！这里有着多么高的智慧！看看锡耶纳的坎波广场、威尼斯的圣马可广场和罗马的人民广场（图35），它们是如此巧妙地在"自组织"的城市区域集中"切"出一些共同的空间，以及安排周边的建筑在面对这些共同空间时的姿态。我们再看看两个同时代的城市设计杰作。豪斯曼的巴黎改造（图36），在它那看似如此蛮横和单一价值观的街道结构框架下，每个地块里却仍然有着这么多的自由空间，让既有的和新加的产权空间里的主体仍在自由地博弈着，对他们规定只在于面对外部街道的共同界面。纽约曼哈顿的街道结构更加严整（图 37），不仅街块被整齐划一地确定，街块内部的地块开间也如此平均，但是地块的合并、拆分却无处不在，地块的建设强度、立面风格、建筑高度也在根据各自的情况进行调整。这两个案例都在告诉我们，恰当地确定"自上而下"手段的边界，是可以创造出和谐统一且多姿多彩的城市。

对于创意城区，显然这样的"合作"是非常必要的。从城市管理者和规划者角度，所要做的就是发现城区的发展潜力，引入创意产业，并设置适当性和激励性的空间规则，来让既有功能和人群拥抱新的产业和人群，合作来发展一个城区，而不是像我们很多不那么成功的创意产业园区一样置换走既有的业主，引来一批新的业主，同时对空间进行目的并不明确的这样那样的改造。很遗憾的是，成功的"合作"，迄今为止，在实际的城区中还是凤毛麟角。

为了证明这个模式的可行性，我们进行了一系列模拟实验。2017 年和 2018 年两个年度，本书作者带领国际国内研究生共 32 人挑选了上海同济大学周边的两个城区进行城市设计。在这里，我们挑战了以往城市设计教学中就结构论结构的思维方式，要求学生认识到创意城区自发更新的核心力量是创意企业和既有建筑的所有者，引导学生从"自下而上的"的"机制"和"自上而下"的"调控"两个视角同时切入，模拟城市的发展，制定规则来引导城市更新的发生。下面的几个方面，是我们教学过程强调的要点。在这里进行罗列，也方便读者在阅读这两个城市设计提案时进行对照。

1.　　根据城区特质和周边资源，进行产业策划
2.　　制定发展的规则，而不是直接制定空间结果
3.　　从利益相关者角度思考问题，激励和规范措施并举
4.　　公共空间作为产业集聚的激发器

The emergence of creative communities has shown that the motivation of sustainable urban development may stem from each individual through their needs and interaction. Meanwhile, our research has proved, that in the development of these urban areas, the "top-down" approaches tend to dominate guidance of space planning, both its advantage and disadvantage play influential roles. On one hand, rigidness, being insensitive to dynamic trends, especially draconian rules of urban planning and administration are adversaries of these urban developments. We shall not forget the violent confrontation during the early period of the development of 798 Art Zone between artists (as renters) and factory owners (as landlords), as well as the spontaneous marches involving the residents of Tian Ti Fang to protest against an official planning. In the cases of Da Fen Village and Zeng Cuo An Village, if it weren't for the fact that they were sheltered in the "urban villages", which are anarchic territories with little regard to the city rules, how could such huge creative clusters have been developed?

On the other hand, should the "top-down" approaches be absent in giving space planning guidelines? Without widening and renovating the shoreline, the waterfront space in the case of Sha Po Wei would always be only a "good-looking place," and it would not become a waterfront as it is today. Creative enterprises in relation to waterfront side are unlikely to rise to the next level. Similarly, Tai Kang Road, which is the southern border of Tian Ti Fang, would still be a noisy and chaotic "street market" today if it were not for the governance of road use and the new planning of the service facilities. Designating it as "creative industries park" in 2008 legalized those changes of functions and buildings and qualified enterprises for protection by laws, which has drawn the former "marginal" development from "underground" to a burst of prosperity. These initiatives, whether they are "space" or "policy" oriented, have undoubtedly made a huge difference, and none of them could succeed only by "bottom-up" approaches.

What is the relationship between "top-down" and "bottom-up" approaches? Is there a possibility to set an appropriate "collaboration" between the two? Let's simplify the problem further, that is, when we attempt to direct and control the development of an urban area using a "top-down" approach, where is the administrative interface of "top-down" influence? By reviewing our vision in the introduction, which states that creative urban areas are an enlightening phenomenon of urban development, we may say, the question above is not on a specific topic regarding the development of creative communities, but can extendedly cover the development of Chinese cities. How can Chinese cities no longer be spatial products of unilateral "top-down" directives, but rather creative spaces created by urban stakeholders who could be granted more power? With this in mind, can we expect cities also benefit from it thus become more distinctive and dynamic, unpredictable and surprising?

In history, manifold wonderful examples of cities suggest the appropriate interfaces of "top-down" approaches' effect, in addition to which, in various sectors of economy, industry, and society, local individual stakeholders and their associations played a crucial role in urban space planning. as a result, this process of urban development is characterized by collaborative governance and management across boundaries and sectors. Either programmed or spontaneous, this collaboration reflects practical wisdom from both top-down and bottom-up initiatives which have shed influence upon these known examples— In the cases of Piazza Campo (Siena), Piazza San Marco (Venice), and Piazza Popolo (Rome) (Fig. 35) , within dense city areas where "self-organization" is playing its daily role, city planners were talented to "cut out" some common spaces, to arrange the surrounding buildings in relation to them, as well as to design postures for these buildings to interact with those common spaces. Let's also look towards two contemporary urban design masterpieces. In Hausmann's Paris planning (Fig. 36) , within the framework of its seemingly insolent and single–value street structure, there is still enough free space in each block for the occupants of existing and future new properties to play with.The streets of New York's Manhattan (Fig. 37) are even more structured. Not only are the street blocks identified in a uniform manner, but also the plots within the streets are equally constantly spaced, nevertheless, the consolidation and subdivision of the plots are everywhere. The building intensity, facade style and building height of the plots are changing according to their respective conditions as well. Both of the cases support that proper collaborative approaches combining "top–down" and "bottom–up" approaches in spatial development will promote a harmonious and colorful city.

For creative urban areas, it is clear that this collaborative solution is necessary. From the perspective of city managers and planners, what needs to be done is to discover the development potential of urban areas, to introduce creative industries, and to set up appropriate and motivating spatial rules to encourage existing functions and people to embrace new industries and people, rather than to replace the existing owners and functions for unclear purposes (like in those partly failed industry parks). Unfortunately, successful collaborative management in actual urban planning has so far been in want.

In order to prove the feasibility of this model, we have carried out a series of simulation experiments. In 2017 and 2018, the research team led a total of 32 international and domestic graduate students to select two urban areas around Tongji University for urban design. Here, we challenge the way of routine urban design which is command–and–control oriented("top–down"), and instruct students to comprehend that the core strength of the spontaneous renewal of creative urban areas is the creative enterprises and the owners of the existing buildings. We inspire students to embark on the research from a broad view, considering both "bottom–up" mechanism and "top–down" spatial means, and go further to establish rules to conduct urban renewal and growth. The following are the main points that we emphasize in our teaching process, which are listed here in order to make it easier for readers to read the urban design proposals.

1. Anticipation of types of future creative industries, based on analysis of characteristics of urban area and resources in surrounding areas
2. Formulation of regulation for urban growth, rather than eventual spatial consequence
3. Focus both on incentive and regulative measures, from perspectives of stakeholders' interest
4. Public space as stimulator for industrial agglomeration

建筑物的最终形态
Form of buildings

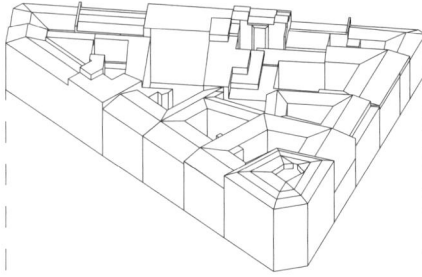

临街建筑必须共同遵守的城市界面
Regulated common urban facade that
related buildings are required to follow

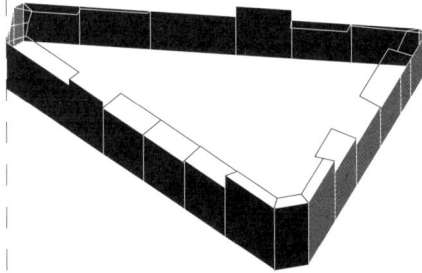

产权地块内建筑物的占地
Occupation of plots by buildings

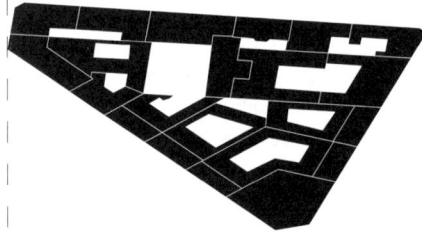

街块和产权地块的组合
Block and combination of various plots

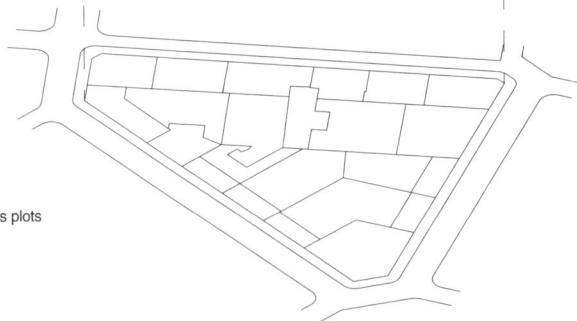

图 36 豪斯曼巴黎改造中典型地块
Fig. 36 Typical urban block in Haussmann's Paris Planning

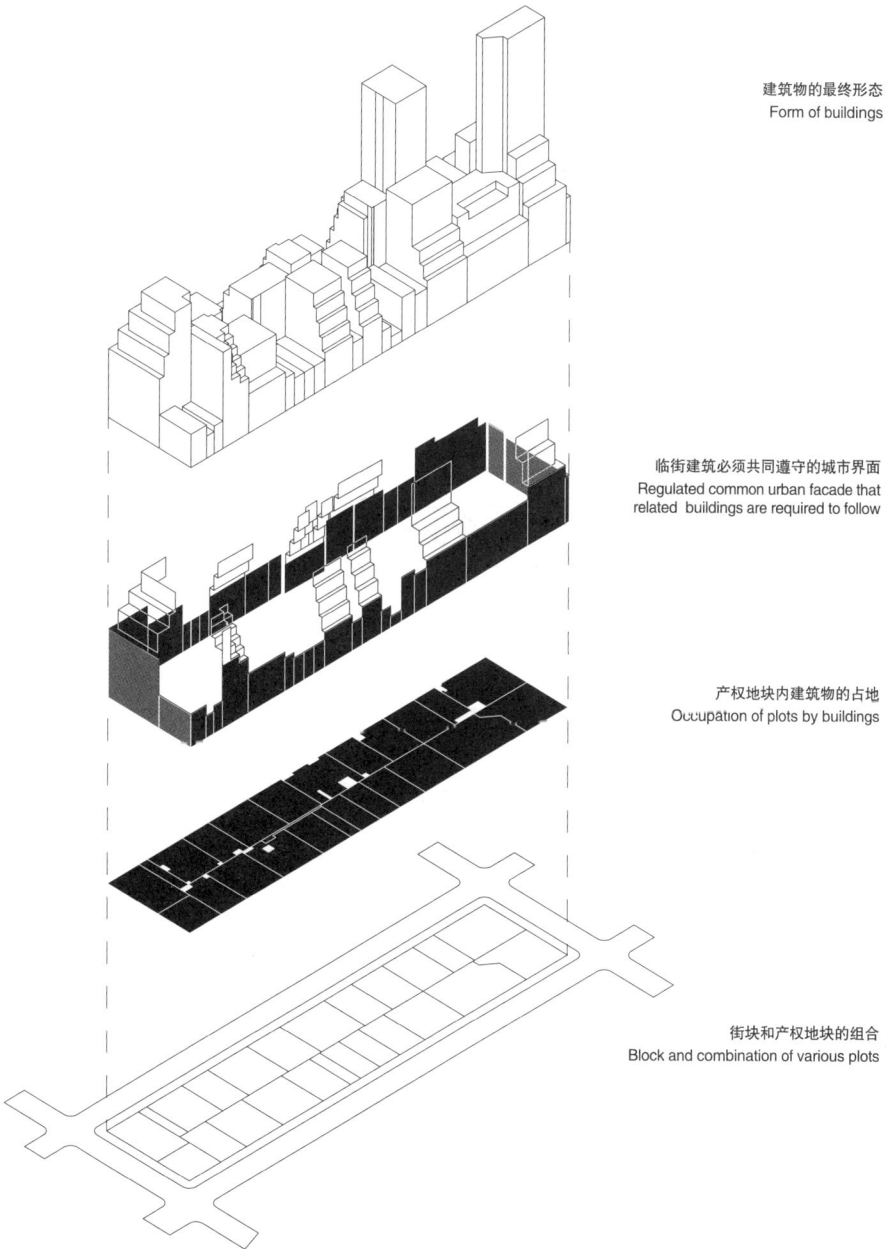

建筑物的最终形态
Form of buildings

临街建筑必须共同遵守的城市界面
Regulated common urban facade that
related buildings are required to follow

产权地块内建筑物的占地
Occupation of plots by buildings

街块和产权地块的组合
Block and combination of various plots

图 37 纽约曼哈顿的典型地块
Fig. 37 Typical urban block in Haussmann's Paris Planning

赤峰路实验

6.2 | EXPERIMENT ON CHIFENG ROAD

赤峰路是同济大学南部边界，长度约 800m。它的一侧是大学校园，另一侧则是由学校、小型工厂、住宅和一些大学的楼宇组成的复杂地块。作为区域的主要交通道路，它承担着东西向的主要交通功能。因为临近大学校园，道路两侧已经零星出现一些培训机构、图文公司和模型公司。我们选择这条街道作为设计对象，既因为在赤峰路两侧拥有发展多样化产业的潜力，还因为现状复杂的产权地块和相应的利益相关者决定了赤峰路的城市更新只能主要通过自下而上的途径来进行。我们的城市设计对该更新过程进行了预测和建议，同时试图建立若干空间用地政策来促发和引导这个过程。

Chifeng Road acts as the southern border of Tongji University, which is about 800 meters long. On one side it is the university campus, and on the other side there is a complex area composed of schools, small factories, residences and some university affiliated buildings. As the main traffic road of the region, it undertakes the main traffic function in the direction of east and west. Because of its proximity to universities, training institutions, graphic companies and model companies have sprung up on both sides of the road. We chose this street as a design object, mainly because of the potential to develop diversified creative industries on both sides of Chifeng Road. In the other hand, the complex property rights and the corresponding stakeholders make it difficult to carry out the regeneration of Chifeng Road through top-down planning methods. Based on the acknowledgement that bottom-up regeneration shall be the main method to regenerate this area, our urban design scheme predicts and suggests the regeneration process, and attempts to establish several spatial policies to stimulate and guide this process.

路网和交通
Road system and traffic

建筑肌理
Figure ground of project area

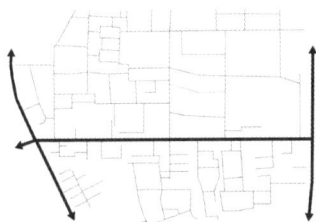

现状步行路径
Current condition

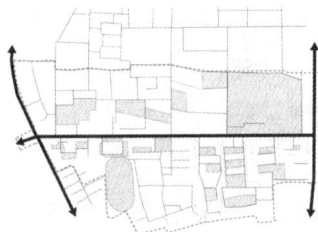

潜在的公共空间
Potential public spaces

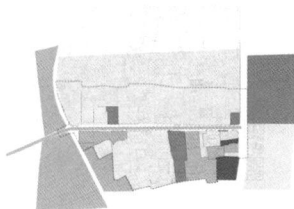

产权划分
Distribution of properties

目前产业
Current distribution of industries

赤峰路两侧的产权状况和相应的利益相关者状况，决定了城市更新的基本途径。当我们怀有这样的愿景，即赤峰路应该成为一条美好的街道，成为这个区域公共空间的主轴，那么我们会看到对于北侧和南侧的发展策略可能会是完全不同。北侧地块的核心问题将是如何开放校园的一部分，让建筑与街道的关系更加紧密；而南侧地块的发展则完全来自于对已有利益相关者的空间关系的梳理，这将带来新的功能和空间的可能性。

The condition of property rights and corresponding stakeholders determines the basic method of urban regeneration. Having the vision that Chifeng Road should become a beautiful street and the backbone of the public space in this area, we will see that the development strategies for its northern and southern sides will be completely different. The core issue of the northern side will be how to open up part of the campus so that college buildings and streets are more closely related, while the development of the southern side will come entirely from regulating the spatial relationships of existing stakeholders. This will open up new function and space possibilities.

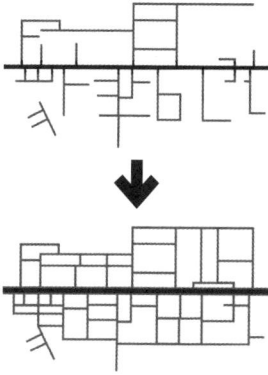

开放与连接内部道路
Open and Connect Internal Pathes within The Plots

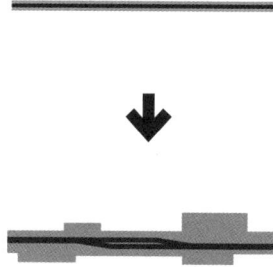

沿赤峰路的公共空间
Public Space along Chifeng Road

向两侧延伸的公共空间网络
Streching-out Public Space

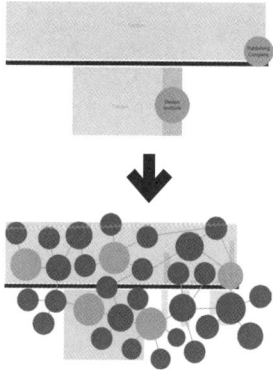

聚合的产业生态
Agglomerated Industrial Eco-system

创意产业对公共空间有巨大的需求，正像我们在案例分析章节所揭示的那样。那么，首先赤峰路的本身应该成为区域里最重要的公共空间。赤峰路上的机动车流量和车速应该被限制，并应该在道路两侧设置更利于行人驻留的、环境舒适的步行空间。其次，两侧地块内部的公共空间网络也非常重要，这就需要通过在一定政策引导下的产权地块之间的合作来实现了。

Creative industries have a huge demand for public space, just as we have revealed in the case study chapter. Then, the public space of Chifeng Road will come from the improvement of the road itself on the one hand, that is, lless traffic flow, pedestrian-friendly sidewalk condition, as well as buildings that provides interface and functions to the street. Certainly, it can be only achieved by top-down planning. On the other hand, the public space network within the areas on Chifeng Road's both sides is also important, which can only be achieved by the cooperation between the property areas under a certain spatial policy.

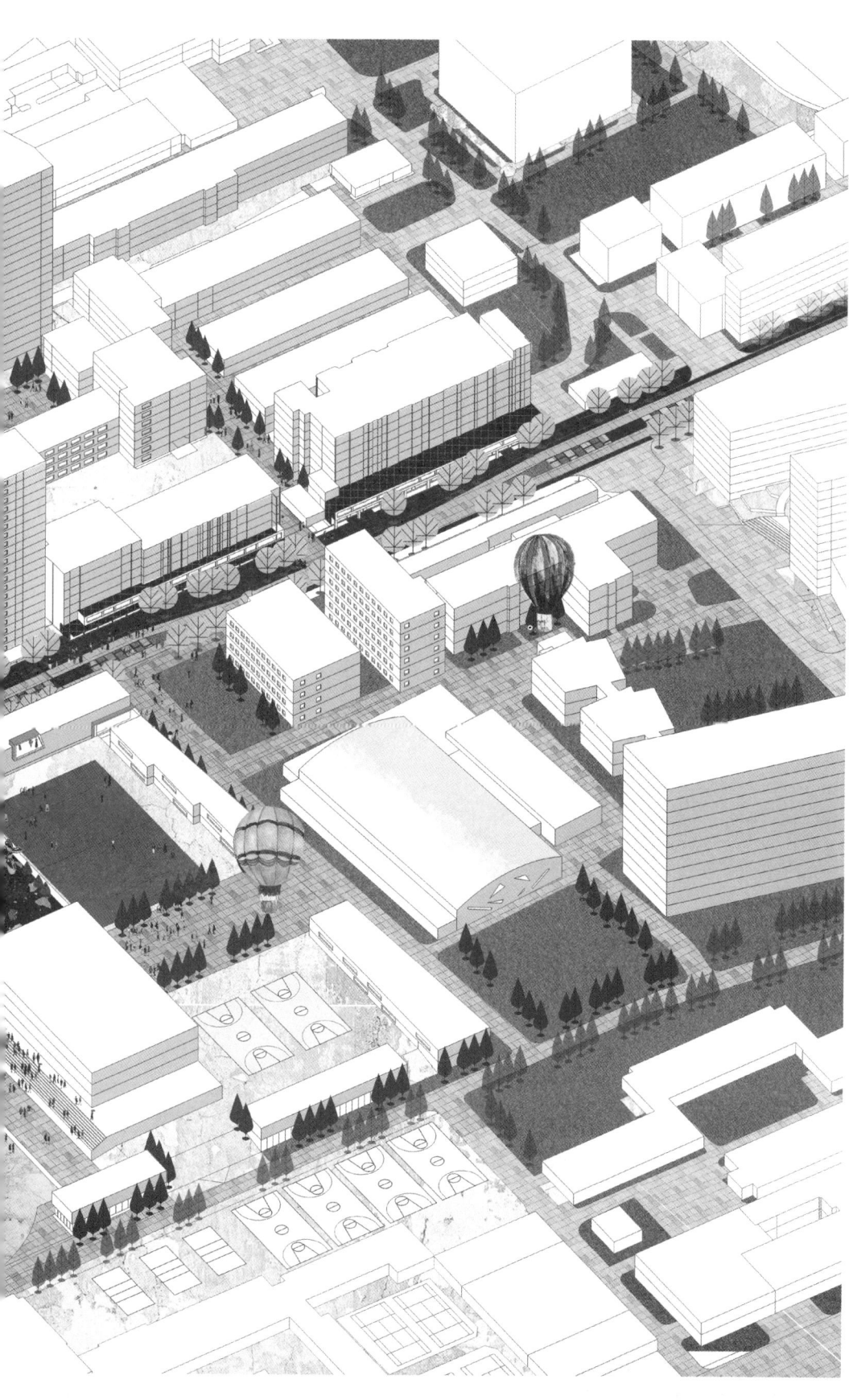

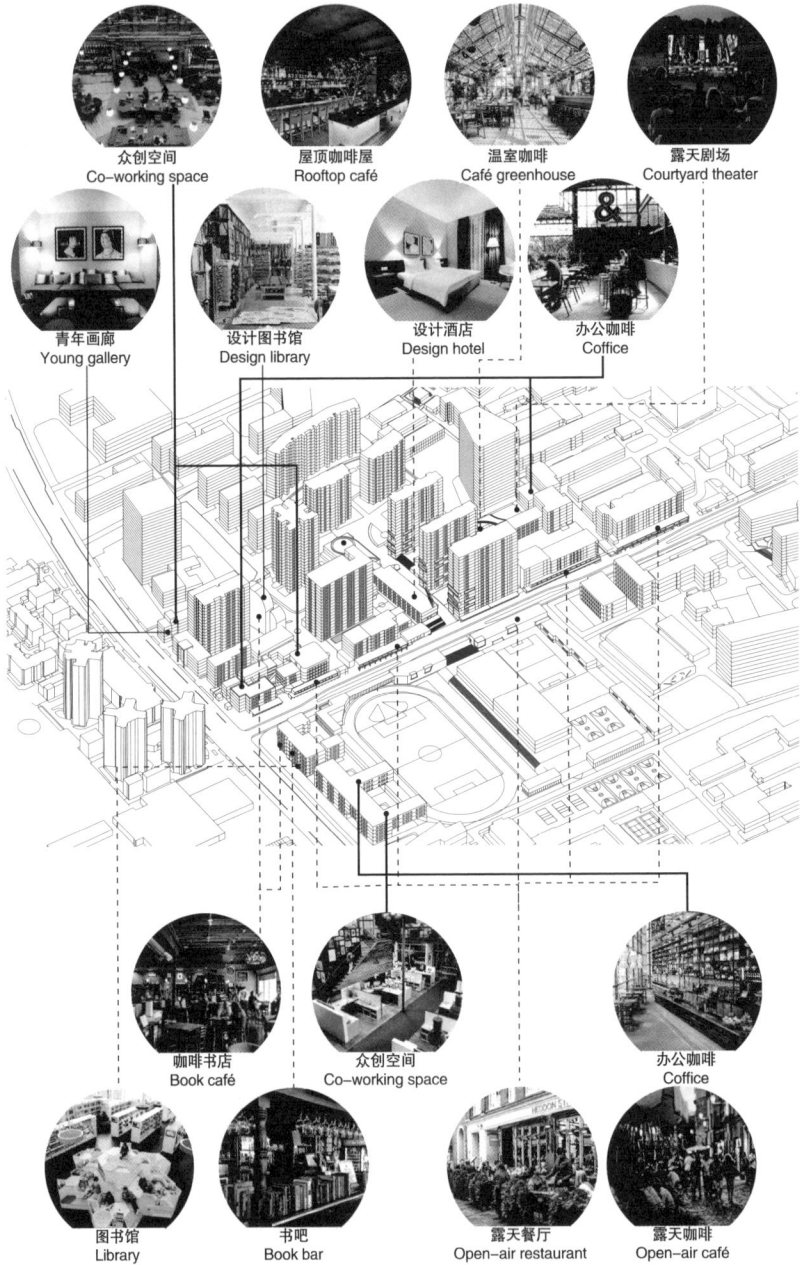

众创空间
Co-working space

屋顶咖啡屋
Rooftop café

温室咖啡
Café greenhouse

露天剧场
Courtyard theater

青年画廊
Young gallery

设计图书馆
Design library

设计酒店
Design hotel

办公咖啡
Coffice

咖啡书店
Book café

众创空间
Co-working space

办公咖啡
Coffice

图书馆
Library

书吧
Book bar

露天餐厅
Open-air restaurant

露天咖啡
Open-air café

可能嵌入的产业
Potential embedded industries

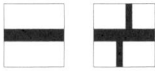

可达性
Accessability

庭院
Courtyards

产业
Industries

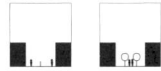

产权
Property

创意产业的聚集需要两个基本条件，一是"创意资源"，它能够为产业发展提供人才、知识和需求。二是符合"空间要素"的特定空间。赤峰路及其两侧的空间，是具备上述条件的，我们认为这里会成为教育和设计产业的集聚区。当然，从事这些产业的创意人员所需求的那些城市服务功能（酒店、特色餐饮和各种城市休闲功能），也会随之集聚。他们将根据既有房屋和城市空间的特质，选择不同的空间类型入驻，有的临近道路，有的深入到街区的内部。随后，产业将从改变这些空间类型开始，逐渐扩展到对公共空间的改造，进而改变整个城区。

The agglomeration of creative industries needs two basic conditions, one is "creative resources", which can provide talents, knowledge and needs for industrial development. The other is the special space according with "qualities of space", which we configure in the chapter 4 of this book. Chifeng Road, with its adjacent urban areas in both sides, is equipped with the above conditions. We believe that it will become ideal area for education and design industry. Of course, the kinds of urban service functions (hotels, specialty restaurants, and a variety of urban leisure functions) that creative people in these industries need will also come together. They will choose different types of space to settle in, based on the characteristics of existing housing and urban space. Some of them will prefer to settle close to the road, and some deep into the inner area of the blocks. Subsequently, creative industries will start from changing these types of space, and gradually expand to the transformation of public space, until they eventually change the entire urban area thoroughly.

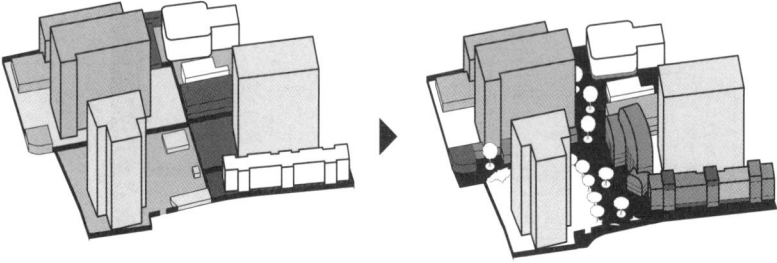

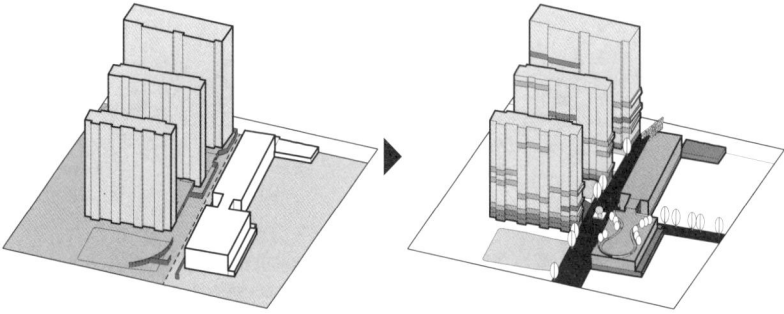

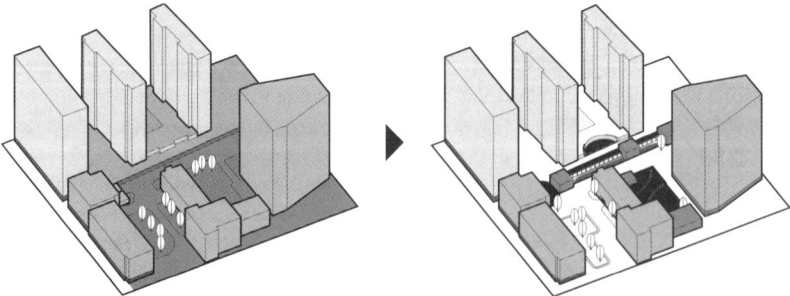

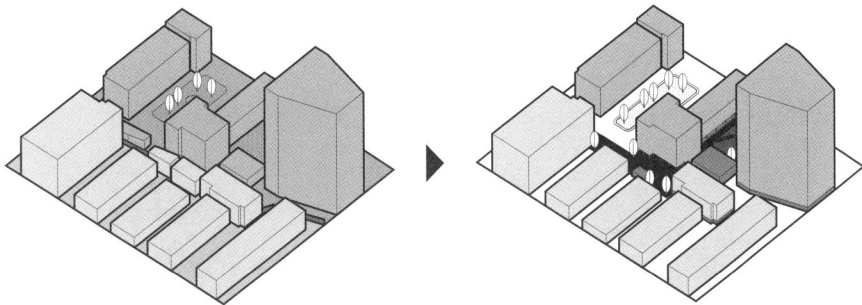

空间类型一：城市走廊

围墙，是目前地块边界的最常用形式，用以保证地块的独立性和安全。而创意产业的需求恰恰相反，是与周边地块和人群的充分接触。只有更开放和更公共，才能给创意产业带来源源不断的活力和创新力。我们的建议是，拆除围墙，让地块和地块之间形成共享的空间，让建筑物直接面对公共空间，让建筑物把他们底层的外墙都打开。底层的空间都变成创意办公、创意商店和各种咖啡厅、餐饮以及体验空间。这些"城市走廊"相互连接，成为覆盖区域的网络，正如同我们在田子坊和曾厝垵里看到的那样。

Spatial type 1: urban corridor

Wall, is the most commonly used form for defining property land boundary, which is to ensure the independence and security of each property. Creative industries, on the other hand, need full access to surrounding plots and people. Being more open and more public, creative industry gains more vitality and innovation. Our proposal is to remove walls, to form shared spaces between property lands, to allow buildings to face directly into the public space and to allow the buildings to open the outer walls of their ground floors. The ground floor spaces are to be transformed into creative offices, creative shops and a variety of coffee, dining and experience spaces. These corridors will connect to each other and form a network covering the area, as we have seen in Tian Zi Fang and Zeng Cuo An.

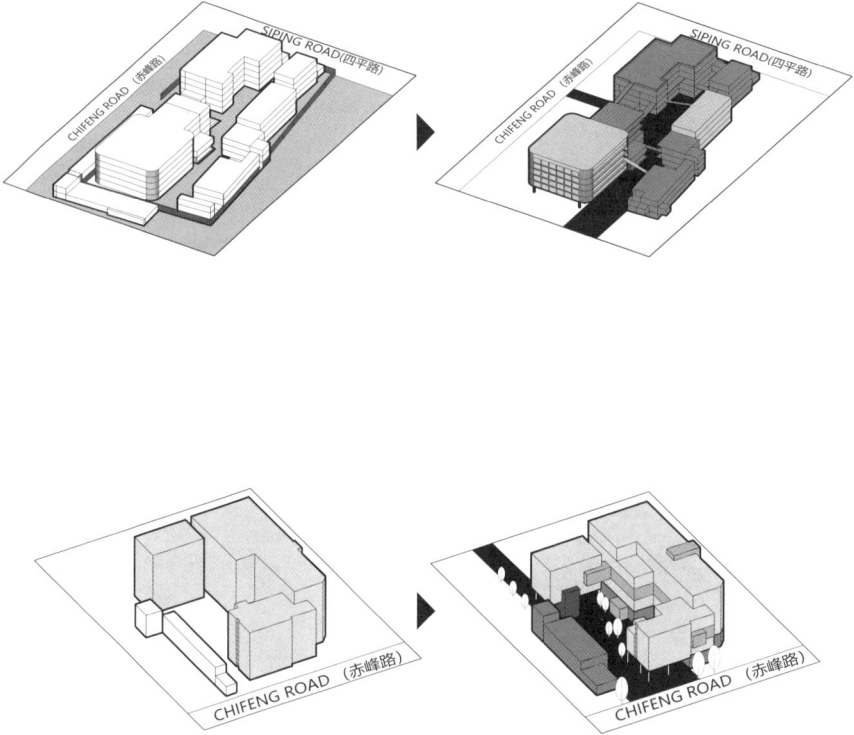

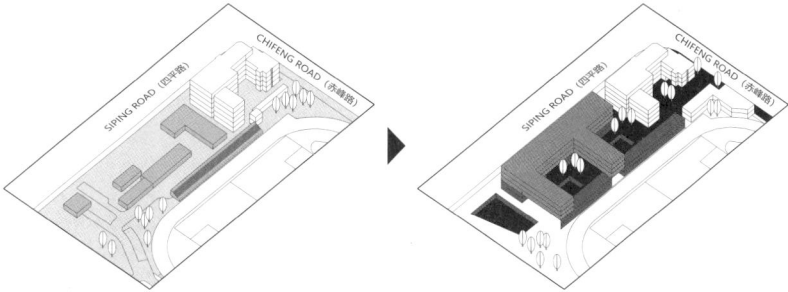

空间类型二：城市庭院

　　有一些空间，可以成为让人驻足和活动的城市庭院空间。他们往往是由本来并不关联的不同建筑单元围合而成。这些庭院空间都是面向公众开放的，拥有不同的主题，成为"城市走廊"形成的网络里的一个个特别的节点。对于区域里的创意产业而言，他们会构思出利用这空间的 N 种方式：创意集市、雕塑广场、户外课堂、都市农庄……意义重大的是，它们的存在让不同的地块结合成整体，创意社区得以形成。

Spatial type 2: urban corridor

There are some spaces that can be used as urban courtyards for people to stop and move around. They are often made up of space between different building units that are formerly not related to each other. These courtyard spaces are open to the public and have different themes, serving as special nodes in the network of "urban corridors". For the creative industries in the area, they will come up with numerous ways to use these spaces: creative bazaars, sculpture plazas, outdoor classrooms or urban farms. What significant is that the existence of these courtyards allows different enterprises and people here to form a whole creative communities.

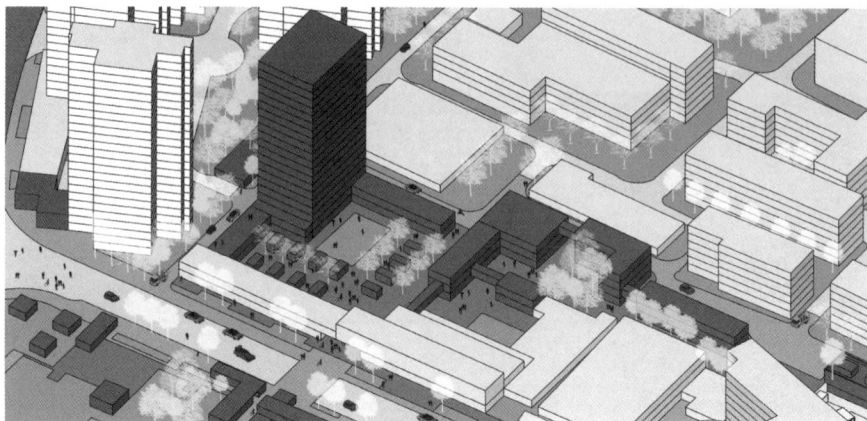

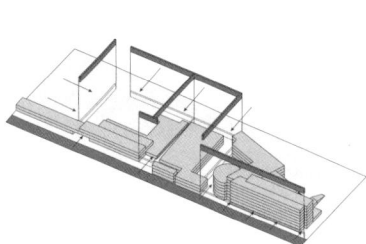

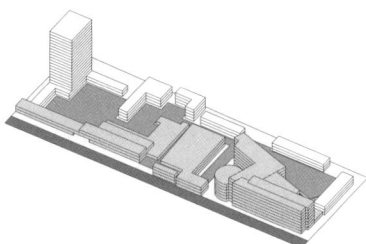

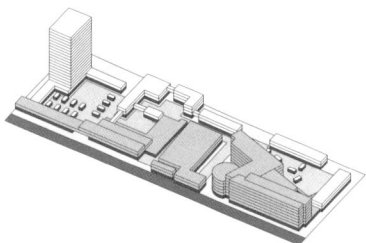

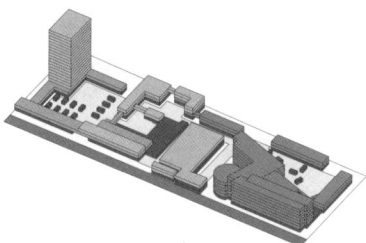

空间类型三：活力岛

Spatial type 3: activity island

赤峰路北侧，进行一定的改造就可以将原本不相关联的校园建筑彼此整合，共同营造出富有活力的公共空间。它是吸引创意产业入驻的重要吸引点。大学的科研资源和产业在这里相遇，相得益彰。

In the north side of Chifeng Road, those formerly unrelated college buildings can be linked by added buildings with office or facility functions. Vigorous public space between them will become attraction for creative enterprises and personnel, who meet the R&D resources provided by universities here.

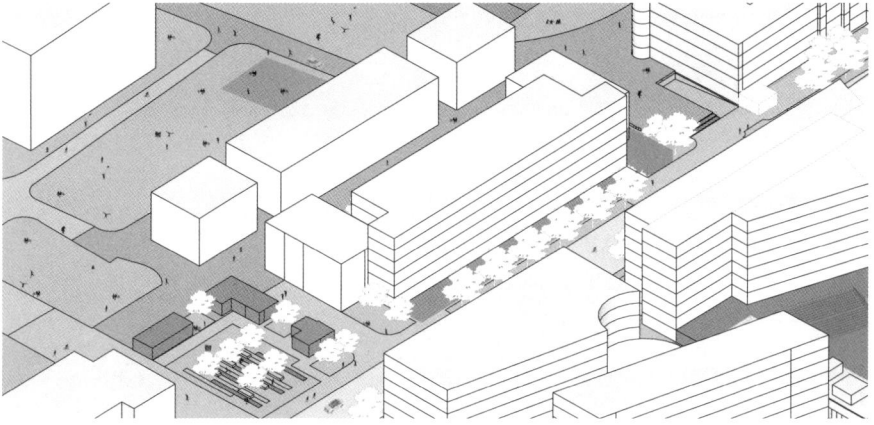

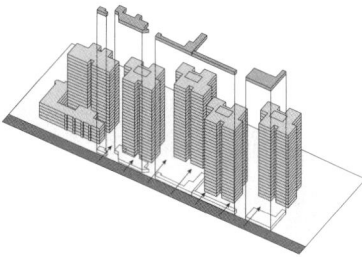

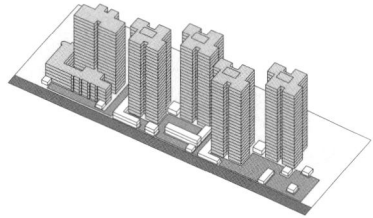

空间类型四：住宅塔楼的 "破墙开店"

住宅区外部的围墙，对街道不是那么友好。更何况对于居民而言，他们也对便利店、洗衣店、儿童游玩设施、餐厅有那么多的需求。那是不是可以把围墙打开，增加这些建筑功能。甚至居住在底层的居民，也允许他们把房屋进行出租，让更多的服务设施可以进入这里。

Spatial type 4: "breaking walls and opening shops" by residential towers

The outer walls of the residential area are not very street-friendly. What's more, for residents, they also have so much demand for convenience stores, laundries, children's playing facilities, and restaurants. Why not do we open the wall and replace them with functional buildings. Even residents, who have their apartment on the ground floor level, are allowed to rent out their homes so that more services can enter the urban area.

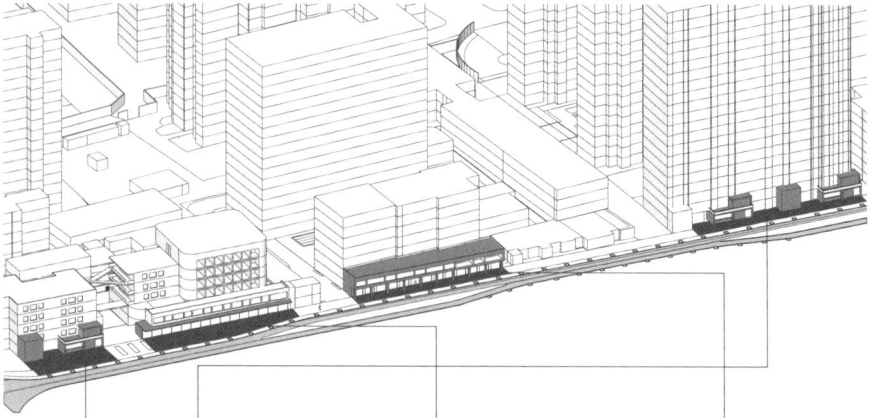

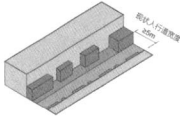

1. 宽阔的人行道空间内设置城市家具。
增加或者切除沿街建筑体量，让界面
更丰富。

1. Wide sidewalk with street furniture.
The building volume is concave and
convex along the street direction.

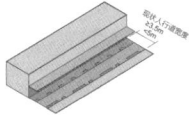

2. 中等尺度的人行道空间。
通过建筑物的出挑（如雨篷）来柔化
建筑界面。

2. Less wide sidewalk.
Use the overhanging method (e.g.canopy)
to soften the building boundary

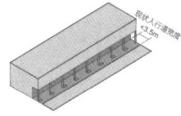

3. 狭窄拥挤的人行道空间。
通过骑楼来扩展人行的空间。

3. Narrow and crowded sidewalk.
Use the arcade floor to expand the
street pedestrian space.

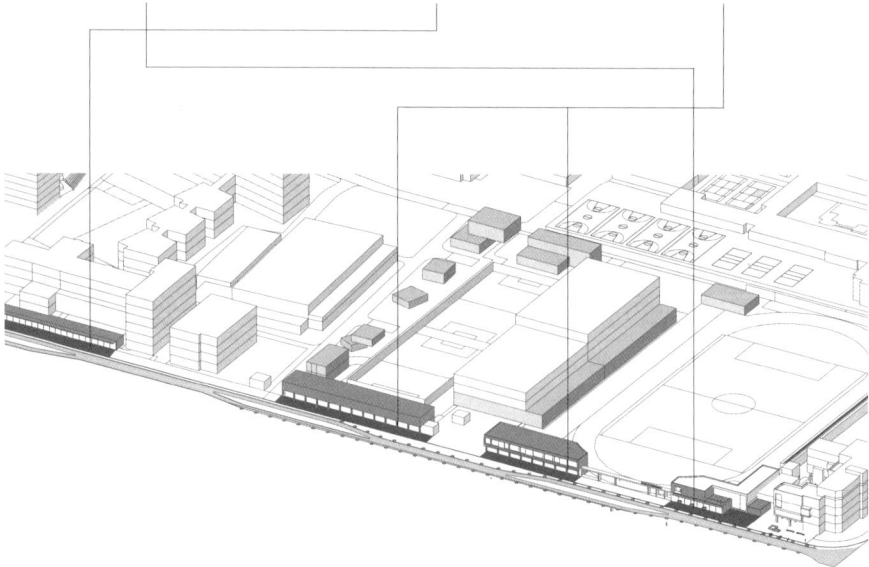

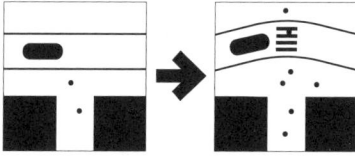

节点
Node

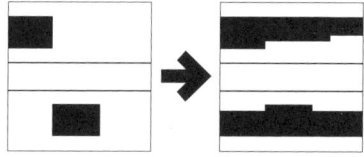

界面
Interface

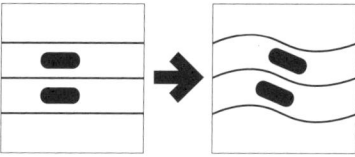

车速
Traffic Speed

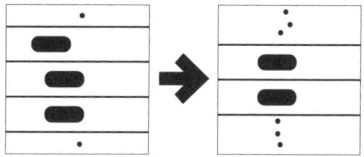

车道
Lanes

赤峰路提升

赤峰路是整个区域的空间主轴，对它的提升，显然很难通过"自发更新"得以实现。这是一个"宏观结构"的问题，这自然是自上而下规划手段的战场了。我们的规划首先是让车子可以慢下来，这需要通过将机动车车道进行拆分和弯曲，有些地方还进行铺装化处理。其次是两侧人行道的环境，让行人不仅可以通行，在他们愿意的时候，还可以停留下来。最后，道路两侧的建筑地块，需要用更连续的建筑界面面对街道（而不是封闭的围墙和隔离绿化带），还须补充大量的商业服务功能。

Upgradation of Chifeng Road

Chifeng Road is the spatial backbone of the entire urban area, which, however, is obviously difficult to be upgraded through the "bottom-up regeneration". This is a typical "macro structure" problem, and this is naturally the battlefield for top-down planning tools. The plan is primarily to slow down the vehicle traffic, by splitting and bending the driveway and, in some cases, paving it. Secondly, sidewalks on both sides shall be improved in term of quality of public space and landscape, so that pedestrians can not only pass, when they are willing, but also stay. Finally, the building blocks on both sides of the road need to face the street with more continuous building interfaces (instead of solid walls and isolation greeneries), and a large number of commercial service functions have to be added to them.

街道改造节点 一
Street reformation 1

街道改造节点 二
Street reformation 2

国康路实验

EXPERIMENT ON GUOKANG ROAD

国康路是同济校园北部的一条街道。它的北侧聚集了很多大规模的设计企业，从事设计行业的人员近万人。对于闻名远近的"环同济产业圈"而言，国康路就是最典型的产业聚集形式。然而，现在看来，它只是一条用来通行和停车的道路。难道这么多创意人员不需要服务设施和美好的公共空间？难道学校和设计企业的地块边界就只能是围墙？当然不是的。我们认为，国康路具有成为一条"美丽的街道"的所有条件，而更新的推动力就应该来自于创意企业和人员对公共空间的需求。我们也相信，国康路的升级更新，会带来两侧产业的进一步发展，不仅在于产业总的体量，更在于其类型的多样化发展。

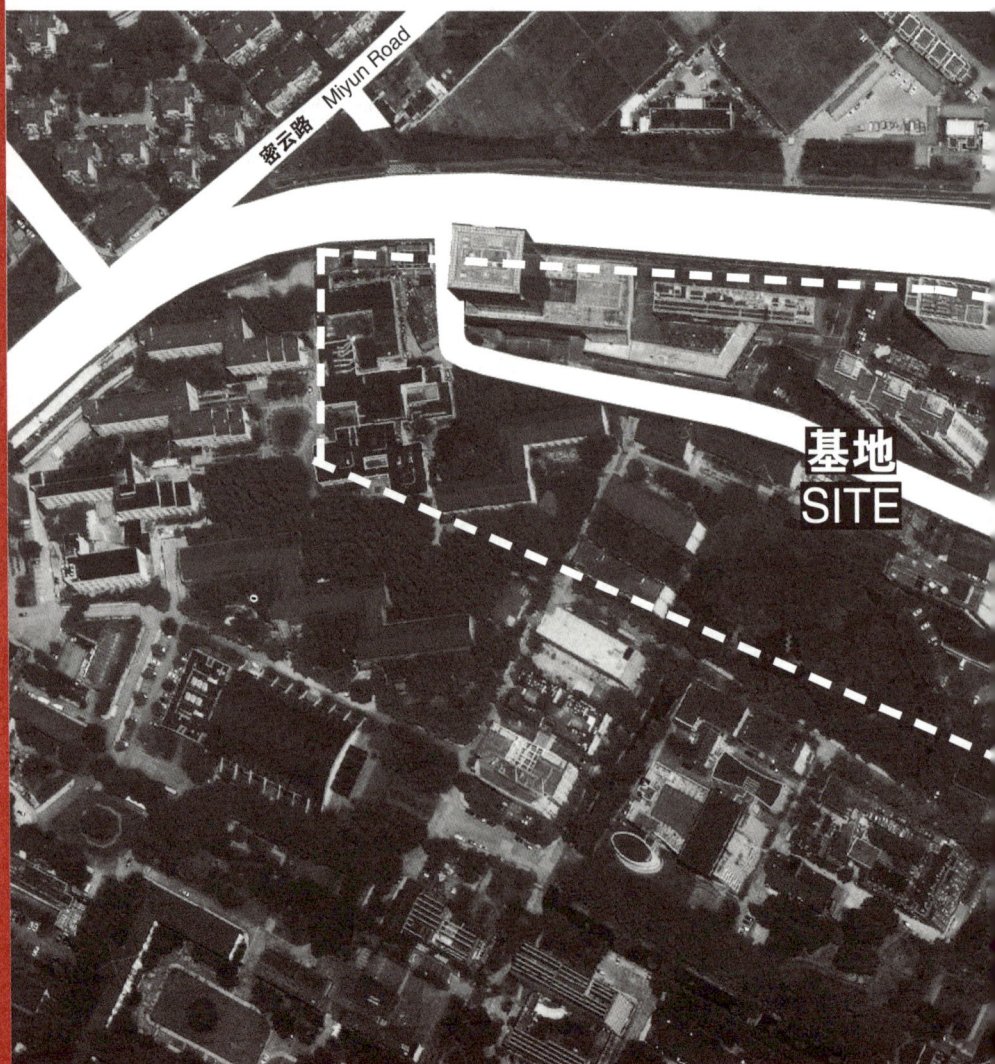

Guokang Road is a street in the north of Tongji Campus. In its northern side, there are so many large-scale design enterprises, in which nearly 10 thousands design profession personnel are working. For the famous "Tongji Industrial Circle", Guokang Road is the most typical form of industrial agglomeration. Now, however, it is just a road for traffic and parking. Don't so many creative people need services and good public space? Are the land boundaries between the school and the design enterprises only in form of fence? Certainly not. We believe that Guokang Road has all the conditions to be a "beautiful street" and that the impetus for its regeneration should come from the demand for public space by creative enterprises and personnel. We also believe that the upgrading and renewal of Guokang Road will bring about the further development of creative industries on both sides, not only in the total volume of industry, but also in the diversification of its types.

中山北二路 Zhongshan North 2nd Road

四平路 Siping Road

国康路 Guokang Road

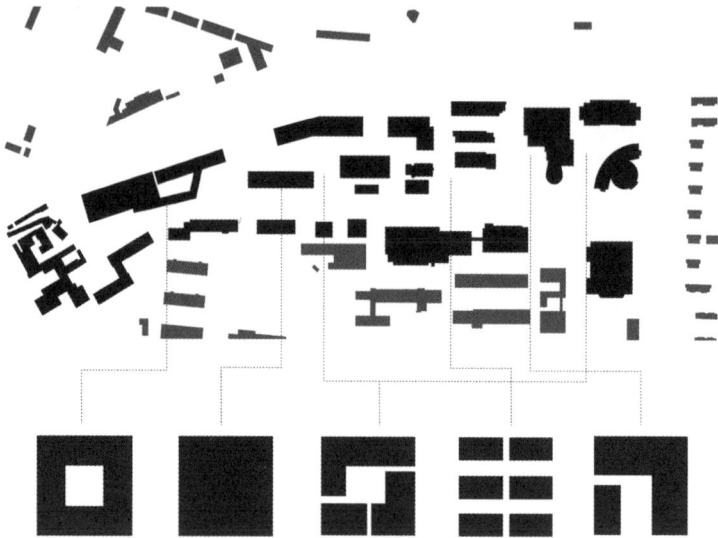

建筑物及其组合的类型
Typology of buildings and their combination

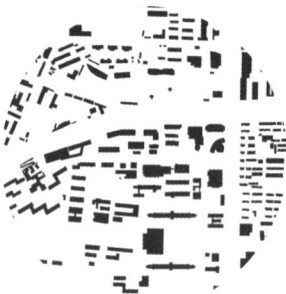

城市肌理
Urban fabric

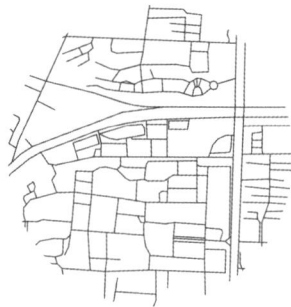

可步行路径
Walkable path

建筑物和空间是双生的。建筑物的组合决定了空间的产生。国康路两侧的建筑物之间，蕴含了那么多的空间可能性。我们首先找到它们，然后研究如何让它们彼此连接。这样，国康路就从一条普通的机动车道路，变成带的公共空间，它串联起两侧的小型化的空间节点和路径网络。新的空间系统，将会成为新的城市活动和新的创新产业的发生地。

The combination of buildings determines the generation of space. Between the buildings on both sides of Guokang Road, there are so many possibilities of space. We find them and then figure out how to connect them to each other. In this way, Guokang Road is turned from an ordinary motor vehicle road, into a belt of public space, it linked up on both sides of the miniaturized space nodes and path network. The new space system will be the ideal site of new urban activities and new innovative industries.

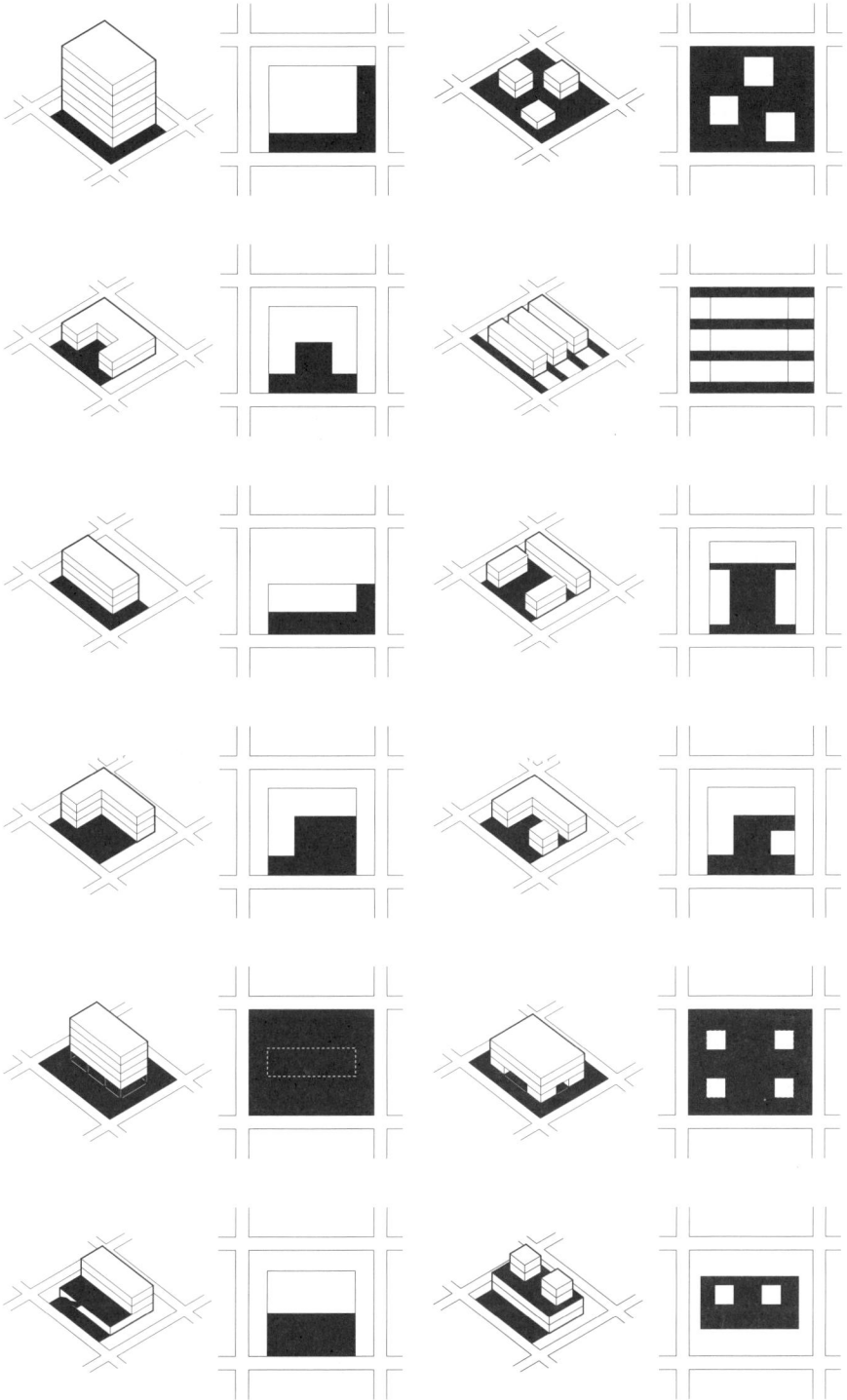

建筑物改建扩建的形态导则
Form-based guidelines for renovation of buildings

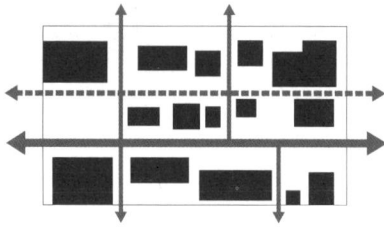

并行的双街体系
Dual Street System

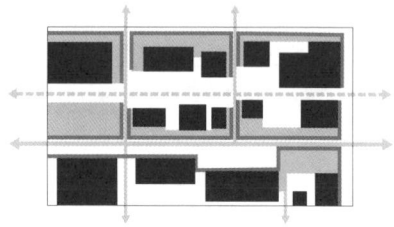

外街对外连续
Continuous Outside

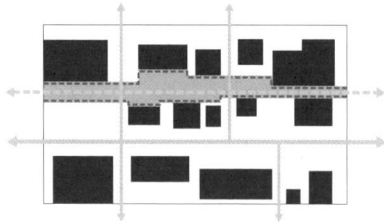

内街对外开放
Open Inside

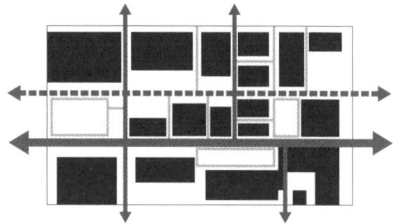

增加建筑以限定公共空间
Create Plaza

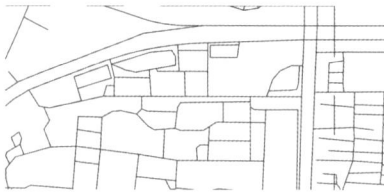

连接可步行路径
Walkable Path

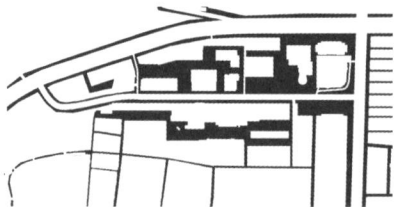

公共空间
Public Space

封闭的独立地块需要被打开，将私有的，也是利用不充分的院内空间开放给公众。这样，地块内的建筑也将会直接面对公共空间。当空间被彼此串联，则会变成平行于国康路的第二条街道。当然，这两条街道是截然不同的。国康路是人车混行的、熙熙攘攘的城市街道，两侧是完整连续的城市界面。而"内街"则是变化丰富的空间序列，它也许不像一条街道，更像是一连串的都市庭院，它成为创意社区里人们日常聚会、交流和举办各种活动的场所。

Those enclosed and detached plots of land need to be more public. Thus the public can access formerly private, and also underutilized space within the compound. In this way, the buildings in the plot will also face the public space directly. When these spaces are connected to each other, they will become a second street parallel to Guokang Road. Of course, the two "streets" are very different. Guokang Road is a bustling urban street lined with people and cars flanked by a complete continuum of urban interfaces. The paralleled "inner street" is a series of urban courtyards, which will become a place for people's daily gathering, communication and a variety of activities.

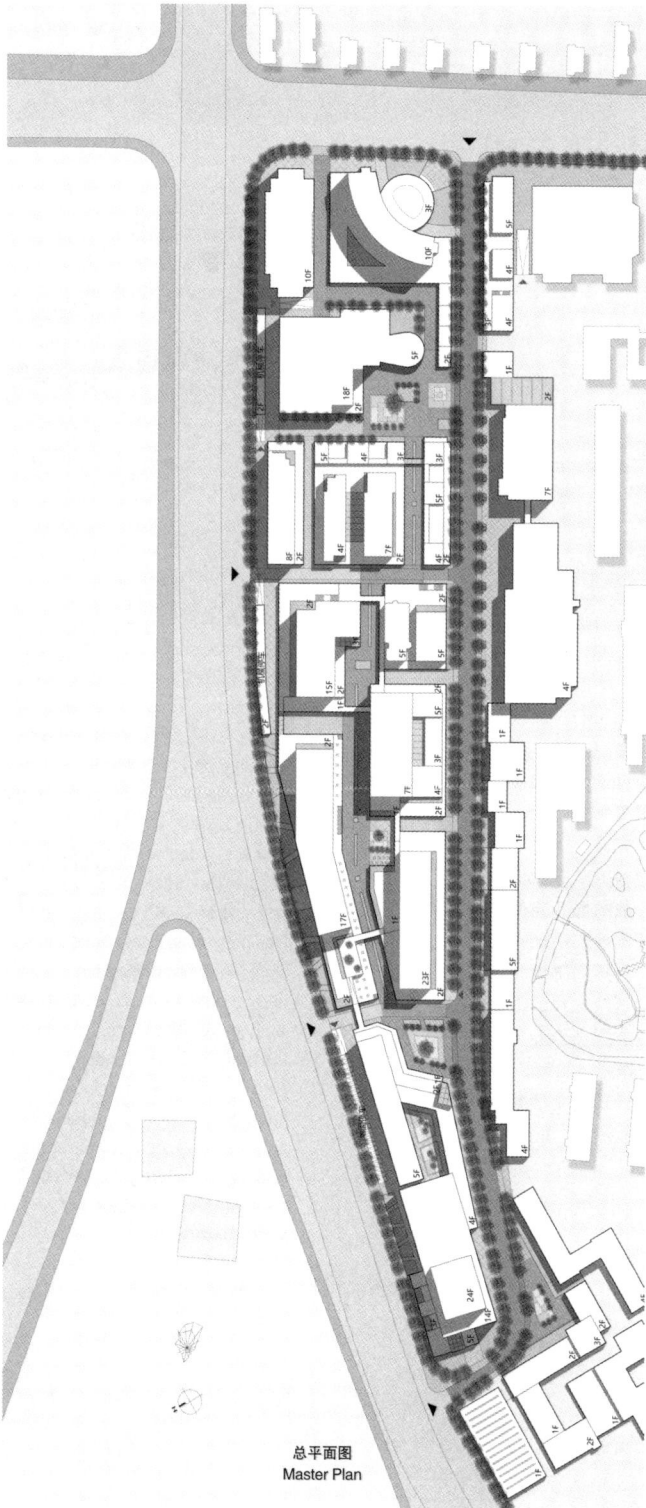

总平面图
Master Plan

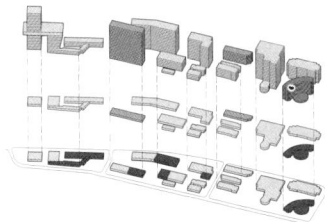

既有建筑的功能改造
New functions for existing buildings

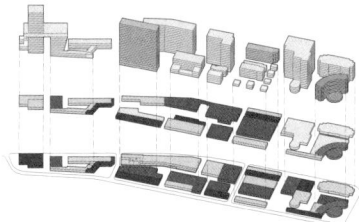

加建建筑和激活公共空间
Adding buildings and activating public spaces

国康路两侧的建筑和规划设计产业，已经形成可观的规模。这和它与同济大学相邻有直接的关系。然而，设计师们难道仅满足于便捷的距离？他们当然会对特色的小店、咖啡厅和餐厅感到兴趣，这些特色空间不仅能服务他们的日常生活需求，也很可能成为设计师们的第二办公空间。在这些共享的、能够让人产生充分交流的空间，创新的能力会得到出乎意料的跃升。设计公司们也会需要一些更加开放的展示空间，来向街上来来往往的人群展示他们的创作成果。让我们再来设想一下，如果同济大学校园的围墙被打开，学生们会喜欢这样的街道。是不是有一些楼房可以变成服务学生的功能，书店、健身房、模型工作间、教室、共享的工作空间，甚至还可以有一些学生宿舍。新的人群、新的产业，在这里将催生新的空间。

The architecture and urban planning design industry related to Guokang Road has formed a considerable scale. This is directly related to its proximity to Tongji University. But are designers content with just convenient distance? They will certainly be interested in specialty shops, cafes and restaurants that not only serve their daily needs, but are also likely to serve as a secondary office space for designers. In those shared spaces that allow people to communicate, the ability to innovate can take an unexpected leap. Design firms will also need more open spaces to show off their creations to the people who come and go on the streets. Let's imagine further, if the walls of Tongji University campus were opened, the students would like this street so much. Can existing and new built buildings be turned into student services, bookstores, gyms, model workshops, classrooms, shared workspaces, and even student dormitories? The new crowd, the new industry in turn, will give birth to the new space here.

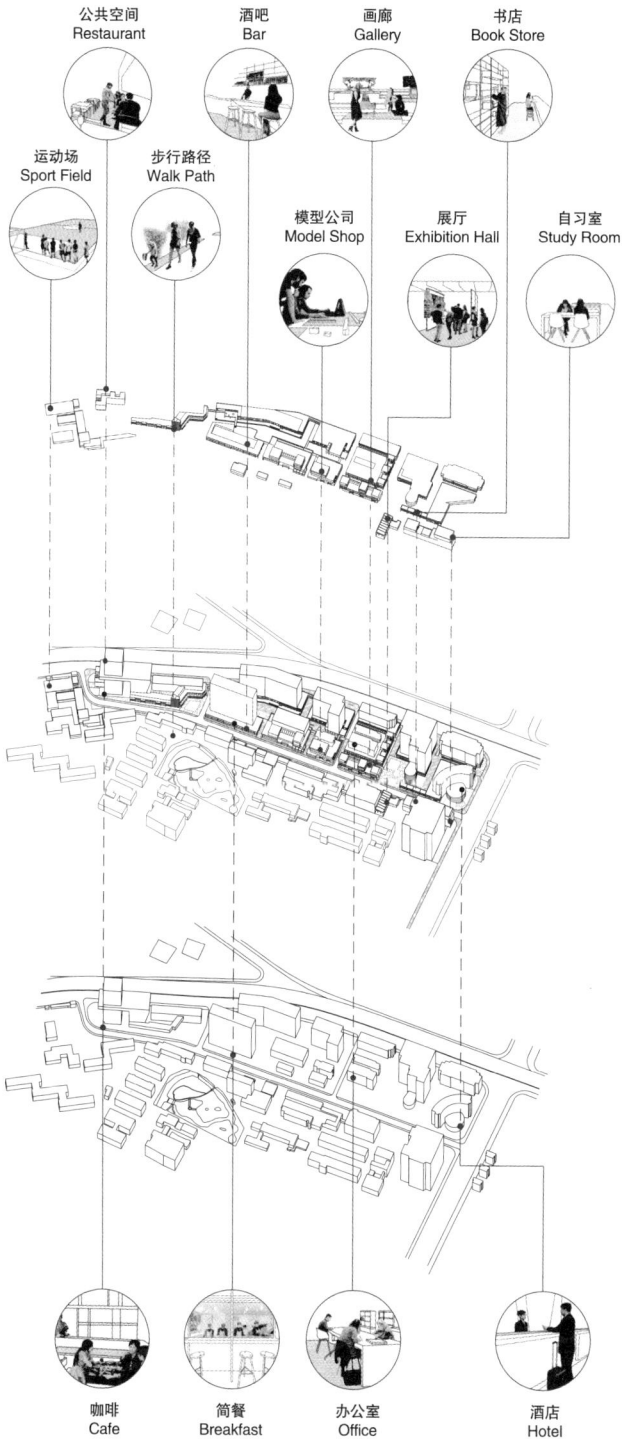

公共空间
Restaurant

酒吧
Bar

画廊
Gallery

书店
Book Store

运动场
Sport Field

步行路径
Walk Path

模型公司
Model Shop

展厅
Exhibition Hall

自习室
Study Room

咖啡
Cafe

简餐
Breakfast

办公室
Office

酒店
Hotel

可能嵌入的产业
Potential embedded industries

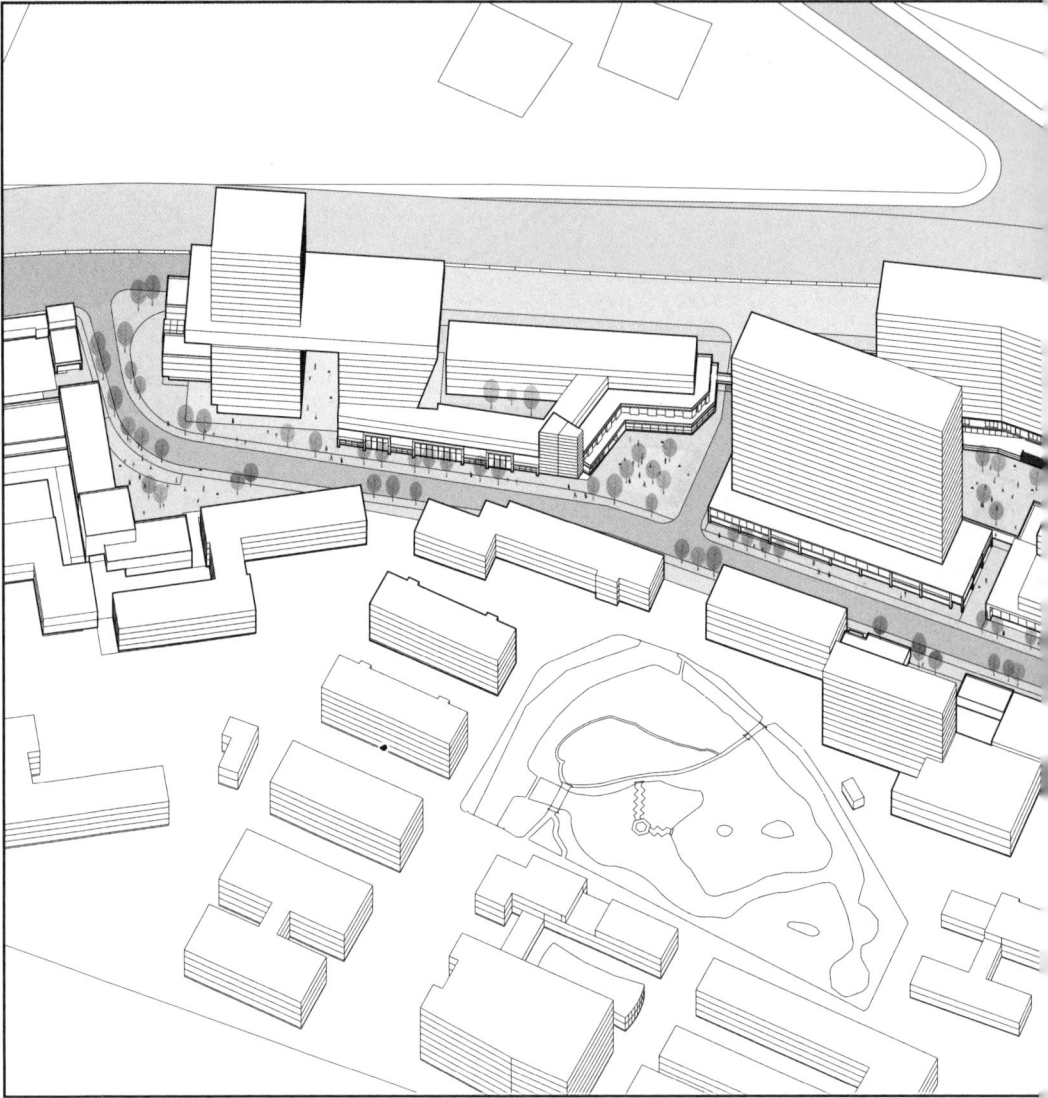

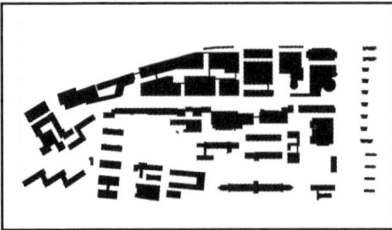

建筑肌理
New urban fabric

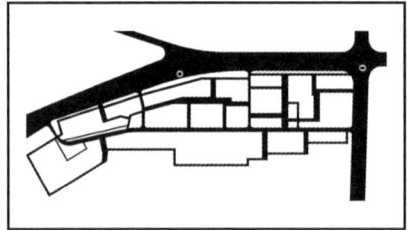

道路结构
New traffic lines

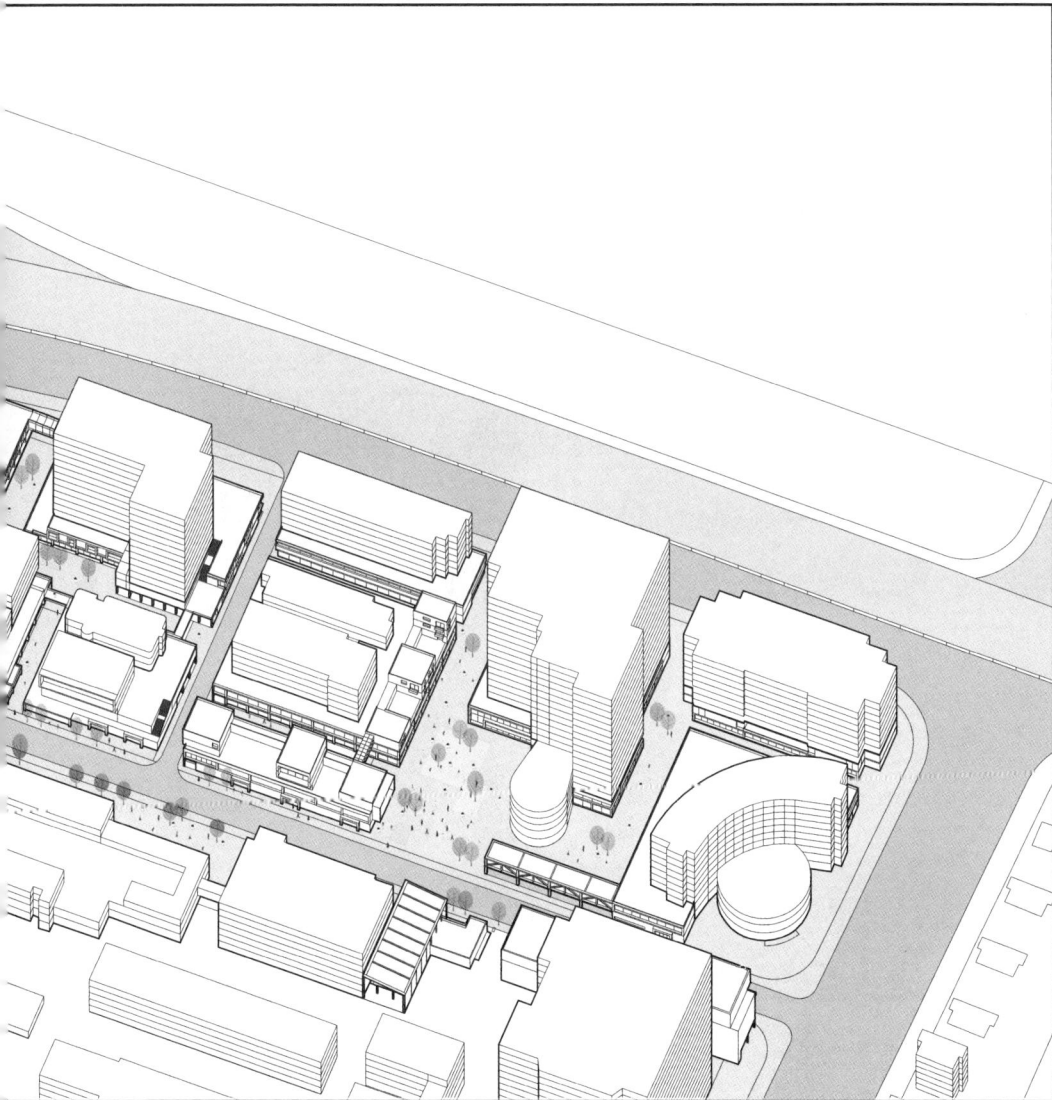

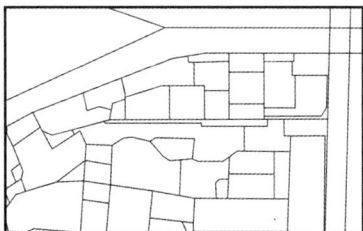

可步行路径
Walkable path

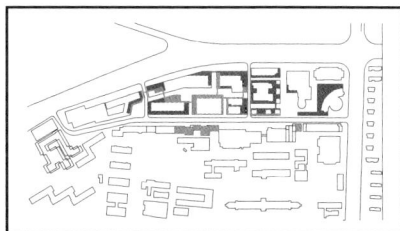

连廊及屋顶花园
Corridors and terraces

内街里的场景 一
Inner street scenery 1

内街里的场景 二
Inner street scenery 2

内街里的场景 三
Inner street scenery 3

内街里的场景 四
Inner street scenery 4

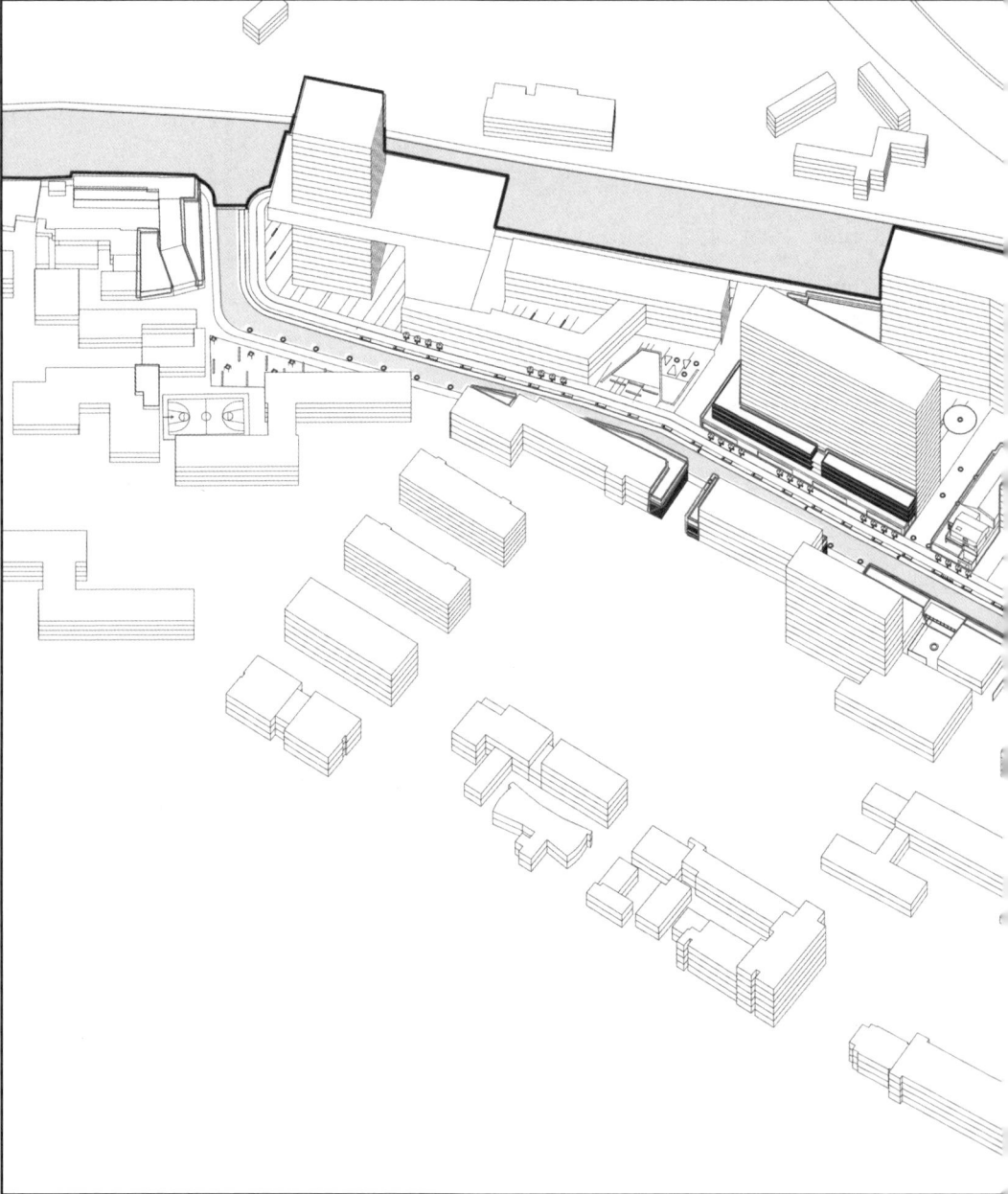

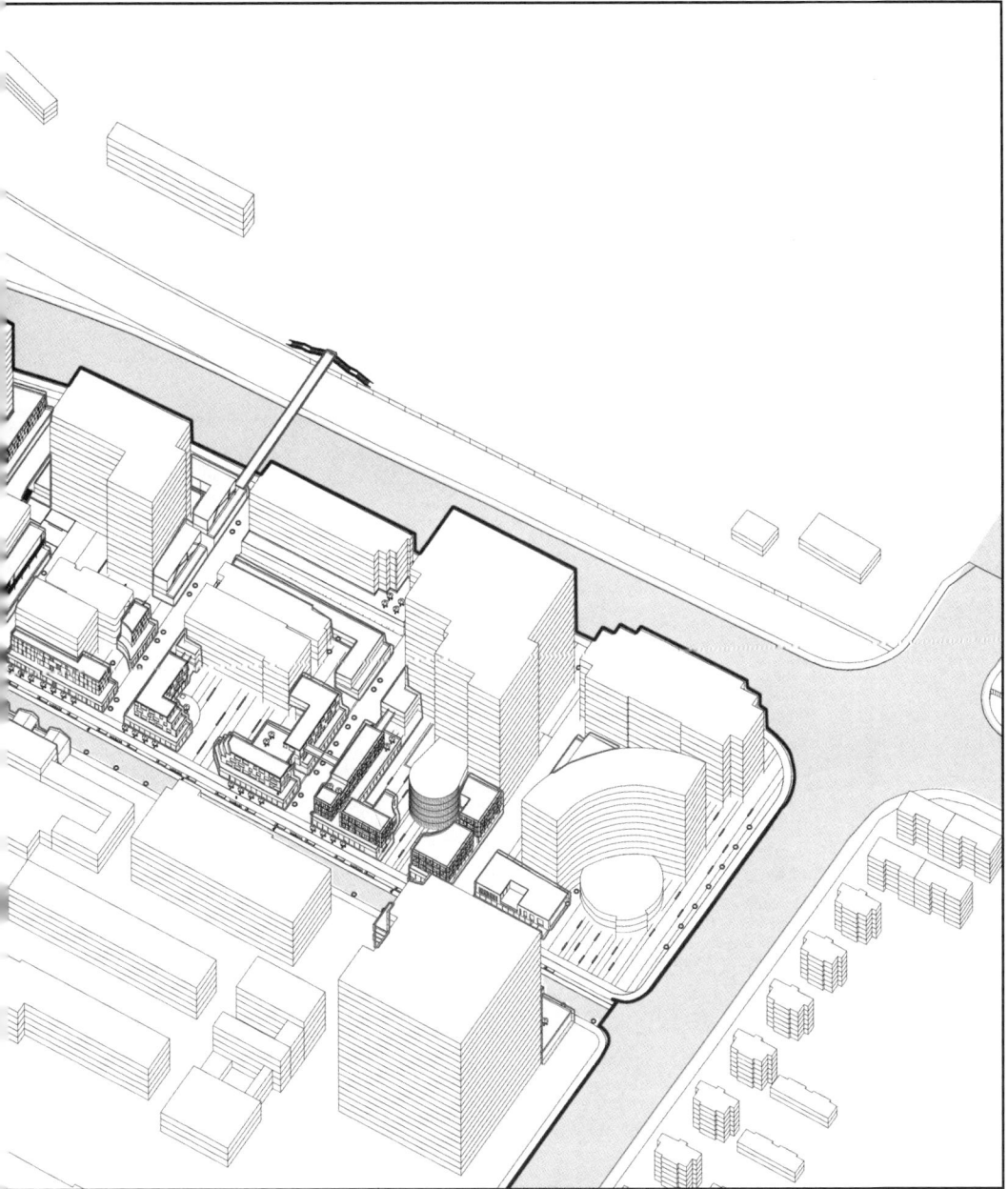

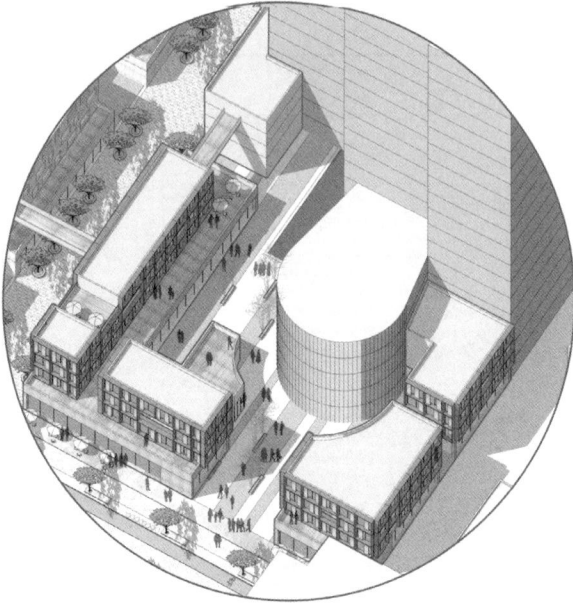

街坊形式 一
Block type 1

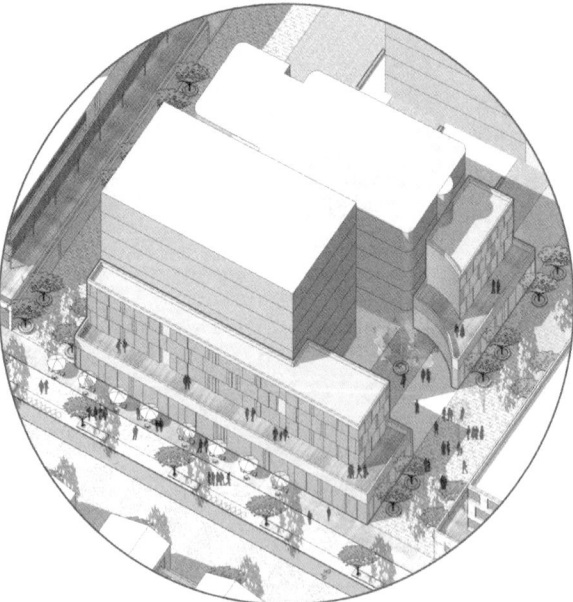

街坊形式 二
Block type 2

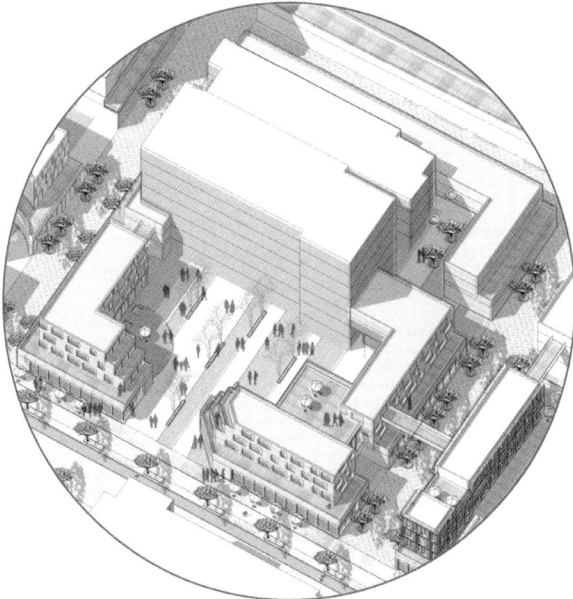

街坊形式 三
Block type 3

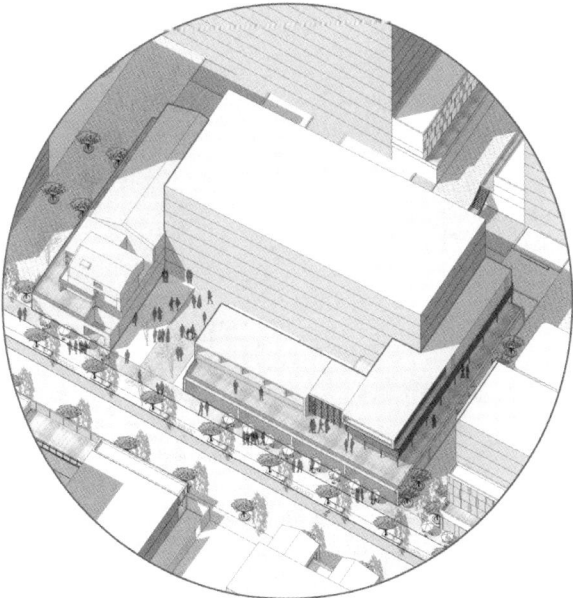

街坊形式 四
Block type 4

关于城市界面的规则
Guidelines for shaping urban interfaces

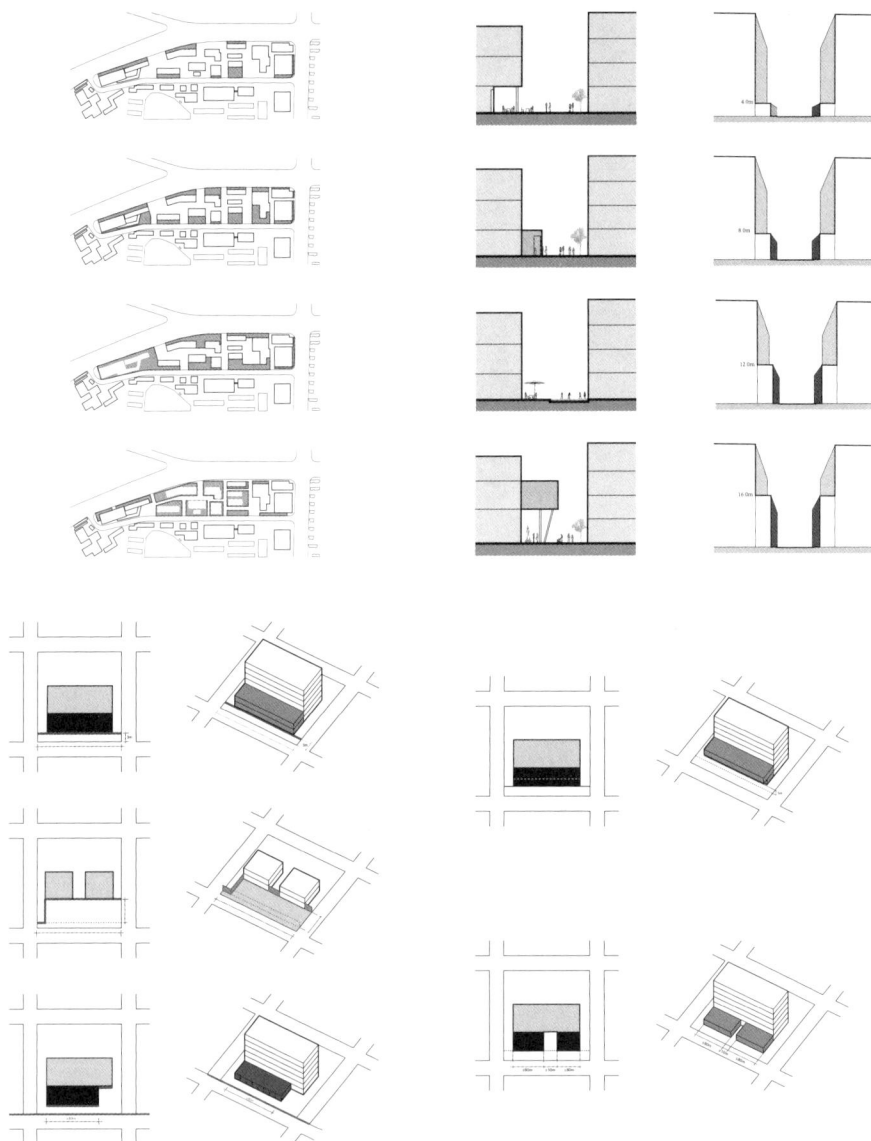

　　城市街道、公共广场和庭院空间，都需要特定的空间形态，而空间形态则需要紧邻空间的建筑体量来限定。城市设计的最终形式，是对城市形态起作用的一系列规则。对于国康路而言，形态的形成需要新的建设，它们必然和已经建成的建筑以及他们的业主息息相关。我们指定的一系列针对公共空间、界面和建筑改建的规则，一方面满足了既有业主的需求（如增加建筑面积和改变建筑用途等），另一方面又必须让他们的新建设符合整个区域未来的空间愿景。这就是"自上而下"和"自下而上"的合作。

建筑加建的规则
Guidelines for building additions

Streets, public squares and public courtyards, they all need specific spatial forms, whose volumes can only be defined by those adjacent buildings. That is why the final form of an urban design is a set of guidelines that play a role in urban forms. As far as Guokang Road is concerned, the formation of the form needs new construction, and they must be closely related to the buildings that have already been built and their owners. We specify a series of guidelines for public spaces, interfaces and building alterations that, on the one hand, meet the needs of existing owners (e.g., increase in floor space and change of building use, etc.) On the other hand, their new construction must be aligned with the future spatial vision of the entire area. In this sense, this is how the "Top-Down" and "Bottom-Up" cooperate.

7

结论：如何规划不可规划之事？
CONCLUSION: HOW TO PLAN THE "UNPLANNABLE"?

在当代城市中，通过"自发"的方式实现发展的城区往往被认为是混乱的、难以管理的和不和谐的。城市创意社区的出现，揭示了在产业主导的情况下，通过"自发"的城市发展途径，可以实现非常具有特色的和活力的城市区域。这也似乎为城市更新找到了一种新的模式，这种模式的影响和潜力值得深入研究。

自组织理论为我们研究此类城市现象提供了基础理论和分析思路，城市是一个开放的复杂系统，具有自组织的特征，创意产业作为城市的子系统嵌入既有城区，与其他的城市功能相互影响和作用，在竞争和协同中实现动态平衡。以此为基础，我们以实证研究为手段，探索了创意社区在选址、空间发展机制和空间形态方面的一些基本规律。这些规律也反过来验证了自组织理论在分析此类城市现象时的优势，例如对动态空间的结构发展以及对微观空间对整体结构影响的解释和分析。本书的研究结论能够提供一定的理论支撑，帮助我们创造出更多适宜创意产业入驻的空间，或促进已经形成的创意社区进一步发展。

这里引出另一个话题，即在创意社区的自发更新中，"规划"是否是必要的。我们的回答是肯定的。在创意社区发展的过程中，审慎和适时的"规划"能在必要的时候对社区发展进行"结构性的提升"。城市管理者应该充分认识到该规律，对于出现自发更新诉求，或是发展至一定阶段的创意社区，主动响应并进行适当程度的规划干预。但需要注意如下三个问题：

1. 规划干预应关注利益相关者之间的协商机制，从各方利益共赢的角度引导和调控空间资源和城市设施建设，以实现完全由当地居民和其他利益相关者自发更新所难以完成的"结构性调整"，帮助创意社区进一步发展。

2. 对于"中观空间"和"微观空间"（对"中观空间"和"微观空间"的定义，参见 3.3），应该从直接的空间干预，转向充分研究基础上的空间政策制定，引导利益相关者在一定框架下的空间再创造，实现个体利益与整体利益的平衡，同时促发空间的多样性。

3. 无论是对哪个层面空间的干预，规划响应都应该体现"实时性"（just in time）和"小尺度"（micro in scale）的特点（尤瓦尔·波图加尔，2012），即对干预的程度和范围严格控制，对实时的反馈非常敏感并根据反馈及时调整规划，以实现通过规划干预对系统发展带来的积极影响。

　　上述结论也同样适用于其他城市建成区域。城市建成区的各类利益相关者相互竞争和协同，是推动城市发展的底层动力。因此，城市是自然生长的，其物质空间的发展呈现出强烈的自组织特征。对这个系统的生长特征和机制，需要长期观察、捕捉和研究分析，而城市管理者对于城市空间的调控，也应该充分重视它自身的特点并善加引导。城市管理者的身份也应该从自上而下的空间管治者，转向游弋于各个空间维度之间的利益协调者。当然，"城市规划"作为一种空间工具，它始终是用来保护公众利益的，这一点永远也不可以有丝毫的改变。

In contemporary cities, the development of urban areas through the "spontaneous" way is extremely rare, they are often considered chaotic, difficult to manage and inharmonious. The emergence of the urban creative community reveals that through industry–led, "spontaneous" urban development, we can accomplish a distinctive and dynamic urban area. It also seems to be a new model for urban regeneration, the impact and potential of which need to be studied in depth.

The self–organization theory inspires analytical thinking for our study and provides an insight into this kind of urban phenomenon on a theoretical basis. In our opinion, the creative industry system is an open and complex system, which is embedded in the existing urban areas and forms a larger system with other urban functions. The development of the system follows the basic laws of the self–organized system. On this basis, we conduct empirical research as a means to explore the basic laws about their location, mechanism of spatial development and forms. These laws, in turn verify the advantages of the self–organization theory in the analysis of the aforesaid urban phenomena, such as the structural development of dynamic space and the interpretation along with analysis of the impact of micro–space on the overall structure. Therefore, the conclusion of this book can theoretically support the idea of creating more suitable space for creative industries or promoting the further development of the formed creative communities.

We also reveal that in developments in many creative communities, cautious and just–in–time "planning" is necessary, which can bring "structural transformation" to the community when it is needed. Managers of the city shall recognize such rule and develop certain planning (or re–planning) when the spontaneity arises in the communities. However, the following three issues need to be considered:

1. Planning should focus on adjustment of "macroscopic structure" in relation to spatial resources and infrastructures, based on full interaction between stakeholders. Such "structural adjustment" is difficult to achieve by self–organization alone.

2. With regard to "mesoscopic space" and "microscopic space", there should be a shift from direct space intervention to space policymaking based on adequate research. Stakeholders shall be guided in a certain framework of space re–creation, so as to achieve the balance between individual interests and public interests, while diversity of space is stimulated.

3. No matter what scope of space it is to intervene in, Planning should reflect the characteristics of "just in time" and "small–scale" (J Portugali, 2012). Here we borrow Portugali's point of view, which is, strictly control the degree and scope of intervention, be very sensitive to real time feedback, and adjust planning in time according to feedback, in order to realize the possible positive impact on the development of the system through "external directives".

The above conclusion also applies to general urban areas. The interaction among large and small stakeholders in the city is the power to promote the development of urban space. Therefore, the urban space system is a system with strong self–organization characteristics. It is our duty to observe, capture and analyze the mechanism of this system and its generating characteristics, additionally, we should pay full attention to its own characteristics when we manage to guide it in certain ways. The role of city's manager shall be changed from the top–down spatial dictator, to a coordinator of public interests between from the bottom and from the top, navigating in various spatial dimensions. Of course, Urban Planning, as an instrument to adjust urban space, must be utilized to protect the public interest. This should never be changed at all.

注释
ANNOTATION

ⅰ．　这样的例子屡见不鲜。例如上海的永康路，本来是上海历史城区里的
一条普通道路。后来因为很多特色餐饮业的入驻，破墙开店，慢慢形
成氛围，发展为一条非常为特色的街道，甚至在国际上都有一定知名
度。后来因为一些卫生问题以及少数居民的纠纷，有关部门以"违章
建筑"为名，对商铺进行拆除，恢复永康路的原始状态。这让上海痛
失一个具有特色的场所。设想一下，有关部门如果能够采取措施解决
纠纷，制定更好的管理措施，"强拆"应该不是唯一的手段。

Such examples are not uncommon. For example, Yongkang Road
in Shanghai was originally an ordinary road in the historical district of
Shanghai. Later, many featured restaurants and bars settled alongside the
road and slowly transformed the road into a very characteristic streets. It
became well known among Shanghai citizens and even in the international
community it has a certain fame. However, due to some disputes between
the restaurants and a number of residents, the government authorities
came in in 2015 to demolish almost all business, in the name of clearance
of "illegal buildings". This makes Shanghai lose a unique place. Imagine,
if the relevant authorities can take measures to resolve disputes, develop
better management measures, "demolition" should not be the only means.

ⅱ．　"规划不可规划之事"（Plan the unplannable）见于 尤瓦尔·波图
加尔 《自组织与城市》。

The sentence "plan the unplannable" is from Juval Portugali's *Self-
organization and city*.

ⅲ．　SME 是 Small and Medium Enterprise 的缩写，这是欧盟对小微企业的称谓。

SME is an abbreviation of "Small and Medium Enterprise", whose definition
is given by the EU.

ⅳ．　相关论述见理查德·森内特的《公共人的陨落》和《新资本主义的文
化》，以及亨利·列斐伏尔的《日常生活批判》和《空间的生产》等书。

For extending understandings, man shall refer to Richard Sennett's *Fall
of Public Man*, *The Culture of The New Capitalism*, or Henri Lefebvre's *Critique of
Everyday Life* and *The Production of Space*.

ⅴ．　"创意阶层"的提法，最早见于佛罗里达的《城市与创意阶层》
(Florida R., 2014)。与《创意阶层的崛起》（Florida R., 2014)。按
照佛罗里达的定义，创意阶层主要包含两种人群，第一种是"超级创
意核心"（Super-creative core），他们是各行业中的领军人物，这
些行业包括科学、工程、教育、计算机编程、艺术、设计和媒体。另
一种是"创新职业者"，他们通过掌握的以知识为基础的技能创造价
值，他们从事健康、商业、金融、法律和教育方面的职业。佛罗里达
还提出，"波希米亚人"（Bohemians）这类人群也是创意阶层的组

成部分（Florida R., 2014），"波希米亚人"这个特定称谓指的是那些边缘的、迁移性很强的流浪艺术家或独立艺术家。例如北京798案例中的主要人群即此，厦门曾厝垵早期的发展也是由该人群推动的。

The definition of "Creative class" is given by Richard Florida in his books *Cities and the creative class* (Florida R. 2005) and *The rise of the Creative Class* (Florida R. 2014). According to this definition, the creative class consists of two main groups of people, the first is the "super creative core" , who are leading power in the industries including science, engineering, education, computer programming, art, design and media. The other is the "innovator", who creates value by acquiring knowledge-based skills and who work in health, business, finance, law and education. Florida also suggests that people like the "bohemians" are part of the creative class (Florida R., 2014). The specific term "bohemian" refers to marginal, migratory street artists or independent artists. For example, the early development of Zeng Cuo An in Xiamen, or 798 Art Zone in Beijing were both also driven by this group of people.

ⅵ. 798艺术区内的主要道路本可以成为较好的线性公共空间，串联几个主要的空间节点。但由于缺乏整体规划，这些道路目前成为停车空间，质量也较差，加剧区内各个节点的割裂。而对这个问题的解决，通过区内企业自下而上的改造是很难实现的。相似问题也出现于深圳大芬村案例。

The main road in the 798 Art Zone could have been a better linear public space, connecting several major spatial nodes. However, due to the lack of overall planning, these roads are now acting as parking spaces, whose quality is also poor, exacerbating the division of the various nodes in the area. To solve this problem, it is very difficult to achieve through the transformation of enterprises in the district by means of bottom-up regeneration. Similar problems arise in the case of Da Fen Village in Shenzhen.

ⅶ. 涨落（Fluctuation）是指物质系统处于热力学平衡态时，作为统计平均值的宏观物理量如能量、压强、分子数密度在其平均值附近有微小变动的现象。又称起伏。涨落是复杂系统的常态运动，系统产生的结构性变化也是通过涨落中的突变实现的。

Fluctuation refers to the phenomenon that the macroscopic physical quantities such as energy, pressure and molecular number density, which are statistical average values, vary slightly near the mean value when the material system is in thermodynamic equilibrium state. Also known as "ups and downs", fluctuation is the normal motion of complex system, and the structural change of the system is also realized by the sudden change of fluctuation.

参考文献
REFERENCES

[1] Coase R H. The institutional structure of production[J]. Occasional Papers L. Sch. U. Chi, 1992, 28: 1.

[2] COASE R H. The Problem of Social Cost[J]. Journal of Law and Economics, 1960, 3: 1–44.

[3] Florida R. The rise of the creative class--revisited: Revised and Expanded[M]. Basic Books(AZ), 2014.

[4] Florida R. Bohemia and economic geography [J]. Journal of Economic Geography, 2.1 (2002): 55–71.

[5] Florida R. Cities and the creative class [M]. Routledge, 2005.

[6] Krugman P. Increasing returns and economic geography[J]. Journal of Political Economy, 1991, 99(3): 483–499.

[7] Marshall A. Marshall M P. The economics of industry[M]. Macmillan and Company, 1920.

[8] Portugali J, Benenson I, Omer I. Spatial cognitive dissonance and sociospatial emergence in a selforganizing city[J]. Environment and Planning B: Planning and Design, 1997, 24(2): 263–285.

[9] Portugali J. Self- organization and the city [M]. Springer Science & Business Media, 2012.

[10] Weber A. Theory of the Location of Industries [M]. University of Chicago Press, 1929.

[11] 孟薇, 钱省三. 建设产业生态学的中国学派 [J]. 科学学研究, 2006(S1):50– 54.

（MENG Wei. QIAN Xingsan. Towards Chinese school in industrial ecology[J].Studies in Science of Science, 2006(S1):50–54.）

[12] 王玲. 科技企业集群的自组织性研究 [J]. 中国科技论坛，2005(5):74- 78.

（WANG Ling. Study on self-organization characteristics of technology enterprises clusters[J]. Forum on Science and Technology in China, 2005(5):74-78.）

[13] 许凯, 孙彤宇. 城市更新背景下中国城市街道网络的生成途径 [J]. 新建筑，2017(4): 30-35.

（XU Kai, SUN Tongyu. The formation of urban street network in Chinese cities on the background of urban regeneration, New Architecture, 2017(4):30-35.）

[14] 许凯，Klaus Semsroth. "公共性"的没落到复兴——与欧洲城市公共空间对照下的中国城市公共空间 [J]. 城市规划学刊，2013(3):61-69.

（XU Kai, Klaus Semsroth. Fall and revitalization of "Publicness": Chinese urban space in comparison with European ones[J]. Urban Planning Forum, 2013(03):61-69.）

[15] 许凯 ,Klaus Semsroth. 城市规划在产业空间移位过程中的角色和作用——以伦敦、汉堡、鲁尔区和维也纳为例 [J]. 城市规划学刊，2014(1):71-80.

（XU Kai, Klaus Semsroth. The role and function of urban planning in redistribution of industrial space-London, Hamburg, Ruhr Area and Vienna as examples[J]. Urban Planning Forum, 2014(1):71-80.）

插图版权
CREDITS OF FIGURES

图 1 厦门沙坡尾航拍照片
Fig. 1 Aerial View of Shao Po Wei in City of Xiamen
来源：作者拍摄

图 2 维也纳 WUK 的内院
Fig. 2 Courtyard in WUK of Vienna
来源：赵畅

图 3 维也纳 WUK 内院里的艺术活动
Fig. 3 Art events in WUK of Vienna
来源：赵畅

图 4 WUK 里的仓库空间被转换成服务艺术活动的空间
Fig. 4 Converted warehouse space in WUK for art activities
来源：Mladen Jadric

图 5 维也纳 Stadtbogen 桥洞下入驻的小型产业
Fig. 5 Small industries beneath the railway bridge in Stadtbogen of Vienna
来源：Mladen Jadric

图 6 桥洞空间里的特色餐厅
Fig. 6 Special restaurant under the railway bridge
来源：Mladen Jadric

图 7 米兰的 Zona Tortona 区里被创意产业激活的地块
Fig. 7 Urban block in Milan' s Zona Tortona which is activated by creative industries
来源：赵月僮

图 8 米兰的 Zona Tortona 区里的某多功能特色空间
Fig. 8 A multi-functional space in Zona Tortona
来源：赵月僮

图 9 城中村里发展出来的创意社区：大芬村
Fig 9. Creative community developed from urban village: Da Fen Viallge
来源：作者拍摄

图 10 里弄住宅里发展出来的创意社区：田子坊
Fig 10. Creative community developed from Li Long housing area: Da Fen Viallge
来源：作者拍摄

图 11 渔村发展出来的创意社区：曾厝垵
Fig 11. Creative community developed from fishing village: Zeng Cuo An
来源：作者拍摄

图 12 上海八号桥的一处办公空间
Fig.12 A working space in Bridge 8 of Shanghai
来源：作者拍摄

图 13 上海八号桥的另一处办公空间
Fig. 13 Another working space in Bridge 8 of Shanghai
来源：作者拍摄

图 14 M50 里的一处办公空间
Fig. 14 A working space in M50 of Shanghai
来源：作者拍摄

图 15 上海八号桥里的公共展示空间
Flg. 15 The chared event space in Bridge 8 of Shanghai
来源：作者拍摄

图 16 田子坊里的街巷空间
Fig. 16 Alleys in Tian Zi Fang
来源：作者拍摄

图 17 沙坡尾里的户外艺术空间
Fig. 17 Outdoor Art Area in Sha Po Wei
来源：作者拍摄

图 18 星辰
Fig. 18 Galaxy
来源：网络开放资源图片

图 19 雁阵
Fig 19. Wild goose array
来源：网络开放资源图片

图 20 蚁巢
Fig.20 Nest of ants
来源：网络开放资源图片

图 21 Nolli 地图的局部（罗马）
Fig.21 Part of Nolli Map （Roma）
来源：Wikipedia 开放资源图片

图 22 火车轨道的转接区
Fig. 22 Shifting area of rails
来源：网络开放资源图片

图 23 贝纳德对流花纹
Fig. 23 Benard cells
来源：Wikipedia 开放资源图片

图 24 一些欧洲、伊斯兰和亚洲城市的肌理
Fig. 24 Urban Fabric of some European，Islamic and Asian cities
来源：作者绘制

图 25 联排式空间单元和空间发展模式
Fig.25 Row units and their development model
来源：作者绘制

图 26 小型独立空间单元及其发展模式
Fig.26 Small individual units and its development model
来源：作者绘制

图 27 大型独立空间单元及其发展模式
Fig.27 Large individual units and its development model
来源：作者绘制

图 28 利益相关者谱系及其与空间形态的关系
Fig.28 Pedigree of Stakeholders and its influence on spatial form
来源：作者绘制

图 29 沙坡尾、曾厝垵、798 艺术区和大芬村的空间结构演进
Fig. 29 Spatial evolution of Sha Po Wei, Zeng Cuo An, 798 Art Zone and Da Fen Village
来源：作者绘制

图 30 沙坡尾、曾厝垵、798 艺术区和大芬村的产业空间布局
Fig. 30 Distribution of industries in Sha Po Wei, Zeng Cuo An, 798 Art Zone and Da Fen Village
来源：作者绘制

图 31 沙坡尾里沿街城市住宅的功能再利用和扩张模式
Fig. 31 Function combination and expansion model in street-side apartment housing in the case of Sha Po Wei
来源：作者绘制

图 32 曾厝垵里村民住宅的功能再利用和扩张模式
Fig. 32 Function combination and expansion model in villager houses in the case of Zeng Cuo An
来源：作者绘制

图 33 798 艺术区里连续多跨厂房的功能再利用和扩张模式
Fig. 33 Function combination and expansion model in multiple-span factory buildings in the case of 798 Art Zone
来源：作者绘制

图 34 大芬村里村民住宅的功能再利用和扩张模式
Fig. 34 Function combination and expansion model in in villager houses in the case of Da Fen Village
来源：作者绘制

图 35 坎波广场、圣马可广场和人民广场里城市建筑共同遵守的界面
Fig.35 Regulated urban interface that buildings shall all follow, in Piazza di Campo, Pizza San Marco and Piazza Popolo
来源：作者绘制

图 36 豪斯曼巴黎改造中典型地块
Fig. 36 Typical urban block in Haussmann' s Paris Planning
来源：作者绘制

图 37 纽约曼哈顿的典型地块
Fig. 37 Typical urban block in Haussmann' s Paris Planning
来源：作者绘制

表 1：创意产业选址的空间要素
Tab. 1 Spatial Factors for creative communities to select location
来源：作者绘制

第五章和第六章中所有未标明图号和名称的图片，版权均来自作者及本书研究团队。

后记

　　2011年，我和太太杨寒，在上海田子坊开了一家小店，名字叫"尤根空间"（Jugend Space）。Jugend在德语中是"年轻人"的意思，我们通过举办活动，希望让年轻有趣的灵魂与自由不羁的思想在此相遇、碰撞，从而萌生创意。在数载的光阴里，弄堂曲折的方寸间，我们和各种部门、居民、业主等人与事有着各种交集、往来，有欢欣亦有惆怅。带着这些憧憬与困惑，我们产生了一个想法，把这些人都邀请到尤根空间来，于2013年举办了一场名为"田子坊大家谈"的活动，让大伙儿来聊聊它的"前世今生"。在那次活动中，我们了解到田子坊是如何从少数艺术家和某个居民，甚至某些官员的"异想天开"，变成今天的"集体创新"，以及而后在"商业化"过程中逐渐失去创新的无可奈何。这让我们每个人意识到，田子坊的发展，正是由这里的居民和每一个创业者共同推动的。那么，它的未来，也是每一个"利益相关者"的未来。

　　我的博士论文研究是关于"城市里的产业空间"。产业区位理论里面有一个重要的观点，那就是企业选择什么样的区位，这是企业个体根据自身的经济模式和社会需求来决定的。某种程度上，这是企业的"自由意志"。产业选址曾经越来越远离城市。而今天，当代企业正在越变越小，也越来越依赖于创意，"自由意志"正让企业重新回到城市。除了田子坊，国内还有很多类似的案例，它们往往都会成为城市里最有活力和有趣的地方。把这些案例集结在一起，呈现给读者，并且告诉大家它们为什么会发展成这样，是促使我们编写本书的初心。

　　本书付印之际，我首先感谢我的合作者孙彤宇教授，我们一起在这本书里投入了大量的精力。无数次的讨论甚至是思想的交锋，让这本书的学术思想得以浮现。他对学术问题的开放态度和勇于挑战未知领域的精神，让我钦佩。我始终记得，当决定带领学生进行关于"城市自组织"的研究和实验时，我们都还只是隐约地感觉到未来的方向，却已经开始行动了，这需要多么大的勇气！我感谢我的好朋友，对我来说亦师亦友的Klaus Semsroth教授和Mladen Jadric教授的帮助，与你们的交流永远那么愉快而有益。感谢好友刘刚和姚栋教授，你们给予我很多灵感，也有很多激励。感谢好友张一兵，你投入了那么多的精力来帮我校对英文，而你对一些问题的"反问"，往往让我惊觉，并重新审视自己的思维逻辑。最后，感谢我的太太，感谢你的陪伴和支持，从尤根到这本著作，每一次欣喜和彷徨，都有你在我身边。

2018年11月13日

EPILOGUE

In 2011, my wife Amelie and I opened a small art space called "Jugend Space" in Tian Zi Fang, Shanghai. "Jugend" means "youth" in German. We hope to let the young and interesting souls meet and collide with each other through various events. In a few years, we have all kinds of interactions with various government departments, neighboring residents, house owners and other people. Joy and melancholy have always been coupled with these interactions. In 2013, we had an idea to invited all these people to be participants of a event called "Let's talk about Tian Zi Fang", where they were asked to talk about their versions of Tian Zi Fang Story. We then started to understand how Tian Zi Fang changed from the "whimsical" of a few artists and residents, or even some officials, to today's place of "collective innovation", as well as how it then gradually lost its innovation power in the process of "commercialization". This makes each of us realize that the development of Tian Zi Fang is driven by every resident and every entrepreneur here. And its future, then, is the future of every stakeholder.

My doctoral thesis is on "industrial space in cities". There is an important point of view in the theory of industrial location, that is, the location of industry is decided by the individual enterprise's own economic and social needs. To some extent, this is the "free will" of the enterprise. After having moved away from the city's central areas, it is clear that contemporary enterprises are getting much smaller in scale and more reliant on creativity, and that "free will" is bringing them back to the cities. Besides Tian Zi Fang, there are many similar cases in China, and they often become the most dynamic and interesting places in their cities. Bringing these cases together, presenting them to the reader, and telling why they have developed in their ways, is the initial idea that prompted us to write this book.

As this book goes to press, I would like to begin by thanking my main collaborator for this book, Prof. Sun Tongyu, for the amount of energy we put into this book together. Countless discussions and even debates on ideas have allowed the book's academic concept to emerge. I admire his openness to academic issues and his courage to challenge uncharted territory. I still remember that when we decided to lead the students in the research and experiment of "urban self-organization", we were only vaguely aware of the direction of the future. It took a great deal of courage. I would like to thank my good friends and collaborators, Prof. Klaus Semsroth and Prof. Mladen Jadric, for their great helps and advices. It has always been so pleasant and rewarding to communicate with you. Same gratitude goes to my good friends Prof. Liu Gang and Prof. Yao Dong. Thanks to my friend Zhang Yibing, you put so much energy into helping me proofread my English. Moreover, your "rhetorical questions" often make me surprised and re-examine whether my logic is clear or not. Finally, thanks to my dear wife, for your company and support. From Jugend Space to this book, with every joy and sadness, I have you at my side.

Xu Kai

November 13, 2018

图书在版编目（CIP）数据

创意产业与自发性城市更新：汉英对照 / 许凯，孙
彤宇著. -- 北京：中国建筑工业出版社，2018.12
ISBN 978-7-112-23102-7

Ⅰ. ①创… Ⅱ. ①许… ②孙… Ⅲ. ①文化产业－产
业发展－研究－中国－汉、英②城市规划－研究－中
国－汉、英 Ⅳ. ①G124②TU984.2

中国版本图书馆CIP数据核字(2018)第288708号

责任编辑：滕云飞
责任校对：王　瑞

创意产业与自发性城市更新

许凯　孙彤宇　著

*

中国建筑工业出版社出版、发行（北京海淀三里河路9号）

各地新华书店、建筑书店经销

上海盛通时代印刷有限公司制版

上海盛通时代印刷有限公司印刷

*

开本：787×960毫米　1/16　印张：14¼　字数：332千字

2019年7月第一版　2019年7月第一次印刷

定价：**68.00**元

ISBN 978 - 7 - 112 - 23102 -7

（33179）